NEBRASKA WILD FLOWERS

Robert C. Lommasson

A BISON BOOK

University of Nebraska Press · Lincoln

Contents

Introduction

This book is an attempt to present to the general public representative types of many of the conspicuous wild flowers found in Nebraska. These 260-some plants in no way constitute a flora of Nebraska which would number some 1,700 species of flowering plants.

Included herein are both native and naturalized flowering herbaceous plants exclusive of the grasses and grasslike plants. Among these are plants that have had a long existence here and others which have recently invaded our area. Some are highly cherished and others are considered as common weeds.

Excluded from this book are woody plants, such as trees, shrubs, and woody vines. Many herbaceous plants with inconspicuous flowers have also been excluded.

All members of a family are placed together, thereby indicating those which are closely related. The sequence of families in the book follows the arrangement of plants in a herbarium and in some technical manuals. The genera within a family are alphabetically arranged. An exception occurs in the composite family in which the family is divided into tribes and the genera are in alphabetical sequence in each of the tribes. In most cases the species are presented in alphabetical order within the genus.

A glossary of technical terms follows the text material. At the insistance of the publisher, keys have been provided for the plants which are listed in the book. There is no expectation that they will be of any use to the general reader. In the index the family, genus, and common names all occur in one alphabetical sequence with species indented and listed under each genus.

Plate 1   Cattail
Typha latifolia

Plate 2   Giant bur-reed
Sparganium eurycarpum

Plate 4   Arrowhead
Sagittaria latifolia

Plate 5   Jack-in-the-pulpit
Arisaema triphyllum

Plate 3   Water-plantain
Alisma plantago-aquatica

Plate 6   Day-flower
Commelina communis

Plate 7   Spiderwort
Tradescantia bracteata

Plate 8   Mud-plantain
Heteranthera limosa

Plate 10   Sego lily
Calochortus nuttallii

Plate 11   White fawn lily
Erythronium albidum

Plate 9   Wild onion
Allium canadense var. *lavendulare*

Plate 12 Sand-lily
Leucocrinum montanum *Cheyenne 6/85*

Plate 13 Western red lily
Lilium philadelphicum

Plate 15 False Solomon's seal
Smilacina stellata

Plate 14 Solomon's seal
Polygonatum biflorum

Plate 16 Dwarf white trillium
Trillium nivale

Plate 17 Yucca
Yucca glauca

Plate 19   Yellow star-grass
Hypoxis hirsuta

Plate 18   Death-camas
Zigadenus nuttallii

Plate 20   Blue-eyed grass
Sisyrinchium angustifolium

Plate 21   Coral-root
Corallorhiza maculata

Plate 23   Showy orchis
Orchis spectabilis

Plate 22   Yellow lady-slipper
Cypripedium calceolus

Acknowledgments
and Picture Credits

This book was written at the suggestion and prodding of the late Professor J. E. Weaver over a period of several years. The kind and understanding support of my colleagues and administrators has been appreciated during its preparation.

Special thanks are due Donald N. Pierce for insisting on the feasibility of such an endeavor as this and for initiating the picture-taking project. Extraordinary thanks are due Mrs. E. D. Weider of North Platte who so generously donated her late husband's large collection of transparencies for my use in preparing this book. Acknowledgments for the use of slides are extended to Mr. Selvester Vanderbeek of Mullen, Mr. Earl W. Glandon of Stapleton, Mr. James Carr and Mr. A. E. Smith of Lincoln, Mr. Robert L. Lemaire formerly of Grand Island and Lincoln, Dr. Robert B. Kaul of Lincoln, Miss Doris Gates of Chadron, Rev. Gale Moon of Chadron, Mr. A. F. Werking formerly of Scottsbluff, and Mr. Truman Doyle of Boonsboro, Maryland. Picture credits appear below.

Thanks are extended to Lloyd Teale, Associate Professor of Romance Languages, University of Nebraska, for his enthusiasm for the wild flowers of Nebraska and for his help in the descriptions of certain composites. Grateful thanks are extended to Mrs. Wanda Bates for reading this book in manuscript and making numerous helpful suggestions. Special acknowledgments are made to Mrs. Irma Gillispie, Mrs. Diana Straub, and Miss Ann Beckenhauer for typing the manuscript. Much patient and understanding criticism has come from my wife, Helen, without whose help this book might never have come into being.

The names of the persons who have contributed photographs to this volume and the plates on which their pictures appear are listed below. Pictures on unlisted plates are reproduced from photographs made by the author. The identification and description of the photographs are the sole responsibility of the author, and any errors should not be attributed to the contributors. All pictures have been made from 35mm color transparencies.

Carr: 17, 59, 123, 257.
Doyle: 33, 120.
Gates: 18, 28, 193.
Glandon: 145, 185, 208.
Kaul: 22, 26, 43, 49, 53, 57, 77, 82, 135, 152, 153, 158, 170, 209, 249.
Lemaire: 13, 27, 51, 54, 80, 83, 94, 95, 125, 130, 133, 146, 147, 162, 176, 178, 190, 199, 201, 212, 217, 231, 232, 235, 243, 244, 245, 247, 250, 251, 252.
Moon: 121
Pierce: 9, 14, 74, 75, 98, 114, 173, 234, 248.
Smith: 16, 55, 124, 180, 181, 219.
Vanderbeek: 78, 122.
Weider: 2, 3, 12, 19, 24, 31, 32, 34, 35, 41, 48, 56, 63, 68, 69, 84, 85, 88, 91, 93, 97, 99, 102, 105, 108, 112, 116, 127, 136, 138, 140, 150, 157, 164, 171, 179, 187, 188, 189, 195, 196, 206, 207, 223, 225, 229, 241, 253, 254, 255, 259.
Werking: 1, 30, 42, 66, 71, 79, 90, 100, 104, 126, 128, 160, 172, 184, 216.

NEBRASKA
WILD
FLOWERS

Plant Descriptions

Monocotyledons

Typhaceae
(Cattail family)

Typha latifolia   Cattail (Plate 1)

Marshy areas are often characterized by the cattails that grow in the mud and water. The ribbon-like leaves vary in width but are unmistakable in that they lack a midrib. The flowers individually are unisexual and quite small and are found in two groups. An enormous number of flowers entirely surrounds the flowering stalk. The staminate flowers which produce the pollen are at the top of the stalk and the pistillate flowers cover a lower portion. The seeds develop from the latter and these persist into the fall and winter while the upper part is often lost before midsummer.

Another species, *Typha angustifolia*, whose leaves are narrower, grows with the common cattail in many places in the state. Ordinarily the upper staminate flowers are separated from the lower pistillate flowers by an interval of bare stalk. In cases of hybridization this seems to be one of the first characteristics to show variation. Flowering occurs in June and July throughout the state.

Sparganiaceae
(Bur-reed family)

Sparganium eurycarpum   Giant bur-reed (Plate 2)

The leaves of this aquatic plant are long and narrow and have a definite midrib. The flowers, borne in separate clusters on the

3

elongate, crooked, flowering stalk which grows up out of the water, are of two kinds. The pistillate flowers occur in greenish clusters on the lower part, while the yellowish, pollen-producing flowers form the upper clusters. This is the only species of *Sparganium* in the United States that has two stigmas on each pistil. These emergent plants should be looked for near the surface of the water among other vegetation and will be found in flower in June and in fruit in July.

Alismaceae
(Water-plantain family)
Alisma plantago-aquatica   Water-plantain (Plate 3)

The tiny, white flowers of *Alisma* are borne on many elongate branches of the panicle. The branches arise in whorls at each node of the inflorescence. The flowers have both stamens and carpels. There are usually six stamens and many flattened carpels on a flat receptacle forming a single ring. Most abundant flowering occurs in July.

The leaves are mostly basal and have the general appearance of those of plantain *(Plantago)* with narrow bases and broad, flat, entire blades. This plant thrives on the muddy shores of lakes and ponds and emerges from the water of marshy areas.

Alismaceae
(Water-plantain family)
Sagittaria latifolia   Arrowhead (Plate 4)

An aquatic plant of shallow water throughout the state is known as arrowhead because of the shape of the leaf blade. The flowers, which are borne on variously branched stalks, are of two kinds, the upper ones bearing pollen while the lower flowers are the seed producers. The flowering stalks appear during July, their white flowers being about one inch in diameter. The plants are not dependent on the seed produced each year because they grow for a number of years from an underground rootstock which is the perennial part of the plant.

4

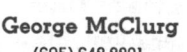
MAY

Sunday	Monday	Tuesday	Wedn

SUNDAY, MAY 17 – JOINT INSTALLATION OF O
POTLUCK TO FOLLOW AT 4:00 P.M.
MONDAY, MAY 25 – MEMORIAL DAY SERVICE (
HOUSE LAWN 9:00 A.M.   POST HOME
FRIDAY, MAY 29 – BLOOD DRIVE 10 A.M. TO 6:

3	4	5	6
	BINGO **6:45 PM**	**POST** **MEETING** **7:00 PM**	**HAMBU** **NIG** **5:00-8:**

2009

esday	Thursday	Friday	Saturday
	4	**5**	**6**
URGER GHT :00 PM			
	11 **AUXILIARY** **MEETING** **7:00 PM**	**12**	**13**
URGER GHT :00 PM			

JUNE

Sunday	Monday	Tuesday	Wed
	1 BINGO 6:45 PM	**2** POST MEETING 7:00 PM	**3** HAMB NI 5:00-8
7	**8** BINGO 6:45 PM	**9**	**10** HAMB NI 5:00-8

2009

sday	Thursday	Friday	Saturday
FICERS WITH N THE COURT- CLOSED 0 P.M.		**1**	**2**
	7	**8**	**9**
RGER HT 00 PM			

Araceae
(Arum family)

Arisaema triphyllum   Jack-in-the-pulpit (Plate 5)

Usually two leaves, each divided into three large smooth leaflets at the tops of tall petioles, arise from a fleshy underground stem. The flowering part arises between the petioles and remains beneath the leaflets. The jack is a spadix. None of the flowers has a perianth, but the whole spadix is surrounded by a special sheath, a spathe, which folds down over the spadix much as in an old-fashioned pulpit with its overhanging reflecting board. By midsummer the spathe has withered away and by late summer the leaves have turned yellow and wilted down. The shiny green and red berries which develop from the pistils are clustered as if at the end of a club.

This plant grows in moist, wooded, humid areas in the eastern part of the state. Flowers occur from April to June.

Commelinaceae
(Spiderwort family)

Commelina communis   Day-flower (Plate 6)

Stems of this dooryard scrambler are erect to decumbent, rooting at the nodes and having lanceolate, clasping leaves. Several flowers are borne in a clasping spathe near the top of the stem. One flower matures each morning. There are two conspicuous blue, roundish petals and a much smaller lower petal which is mostly white. There are usually three fertile and three sterile stamens. The fertile anthers are at the ends of long downward and inward curving purplish filaments while the sterile stamens are conspicuous because of the curious yellow H-shaped sterile anthers. The flower wilts after a few hours and there remains only a drop of blue liquid in place of the petals. Flowering begins in June and may last until September.

5

Commelinaceae
(Spiderwort family)
Tradescantia bracteata   Spiderwort (Plate 7)

There are several species of this genus to be found in Nebraska, all of which go by the same common name. This is the most frequently encountered species, and it ranges throughout the state wherever moderately moist soil is to be found. The bracts of the inflorescence are nearly leaf-like and, in fact, in this species are longer and wider than the long narrow vegetative leaves. The three petals are all alike and are generally pointed. Flower colors in this species as well as in the genus vary from pink to deep blue, although blue would have to be called the predominant color. Old railroad beds may be literally covered with these fascinating flowers during June. *Tradescantia ohiensis* is conspicuous along U.S. Highway 30 from Fremont to North Bend. The six stamens in each flower are fertile and their filaments (thin stalk of the stamen) are covered with purplish or bluish hairs whose cells are large enough to be seen with a 10X hand lens. Under the microscope one may watch the circulation of the cytoplasm within these cells.

Pontederiaceae
(Pickerel-weed family)
Heteranthera limosa   Mud-plantain (Plate 8)

This mud-loving plant produces several to many oblong leaves with sheathing leaf bases. The flowers of this species are produced one at a time and usually last only one day. The flowers are up to one inch in diameter and may be either white or blue. The flower tube is long and is divided into six narrow lobes which turn out abruptly from the tube. Of the three stamens, two have yellow anthers while the other is more greenish and somewhat sagittate. Stamens in flowers of other species are likely to be more distinctly dimorphic than in this species. Flowers occur from July to September. These plants are to be found in such habitats as the marshy basins of Fillmore County where this picture was taken.

6

Liliaceae
(Lily family)
Allium canadense var. *lavendulare*   Wild onion (Plate 9)

The plant pictured is probably the most common variety of the species in the eastern part of the state. Long slender leaves arise from the fibrous covered underground bulbs. The flowering shoots (scapes) may be two feet tall and bear white to pinkish flowers with six thin, pointed perianth segments which mature in natural sequence in contrast to the bulblets that replace the inner flower parts of the wild garlic, var. *canadense.* With age the flowers tend to droop. Flowering occurs during the months of April to June.

Allium nuttallii is perhaps most common in the western half of the state. It is only about one-third of the size of the above species and differs in having thick perianth segments which are rigid in fruit. Flowering occurs during May and June.

Liliaceae
(Lily family)
Calochortus nuttallii   Sego lily (Plate 10)

The beautiful little sego lily is one of two very similar species which are characteristic of the Pine Ridge flora in June and July. It is found in rocky soil among the pines of Dawes County and extends westward into Sioux County.

Although the flowers are usually produced singly on thin, erect stems, they are about three inches in diameter and very attractive. Each of the three white petals has a yellowish base into which is sunken a whitish, moon-shaped glandular area and above is a red brown splotch of color. In this species the anthers of the six stamens have blunt tips and are as long as, or shorter than, the filaments. The ovary matures into a many-seeded, three-sided capsule. The plants are perennial and bear several slender leaves which are from two to six inches long.

7

Liliaceae
(Lily family)

Erythronium albidum   White fawn lily (Plate 11)

Before most people are aware that spring flowers are in bloom the humble little dog-tooth violet (as it is also known) has come and gone, although actual flowering may vary from March until May. These are bulbous plants, as are many lilies, and once a good patch is found they may be located there year after year. These plants occur in thick patches and the bulbs often supply a plentiful amount of food for small burrowing mammals. The splotched, thick fleshy leaves are gracefully arched or, in the early stages of development, folded. The leaves are present as much as two to four weeks ahead of the full flush of blossoms. In moist to dry wooded areas and moist pastures these plants occasionally are found so thick as to appear as a solid carpet. The petals are white to purplish with brownish spots. The three stigma lobes are separate and somewhat recurved in this species. Our most common plant is probably a variety of this species, but the pendent flowers among the mottled leaves are very distinctive.

Liliaceae
(Lily family)   *Cheyenne 6/85*

Leucocrinum montanum   Sand-lily (Plate 12)

Conspicuous clumps of long, narrow, thick leaves wrapped together by thin papery sheaths and arising from the ground characterizes this stemless plant. Underground the short rootstock is supplied from thick, fibrous roots. The sand-lily grows in sandy soil of the western part of the state, and it blooms early in the season from April on into June.

The flowers are sessile and are produced in umbel-like clusters with their flower stalks and the base of the long floral tube underground. There are often four to eight snowy white flowers in a cluster. The similar petals and sepals are called tepals and form a tube as much as three inches long to which the filaments of the six stamens are attached. The anthers of the stamens usually are displayed at the throat of the flower tube. Above the tube the tepals form narrow pointed lobes which turn outward forming a star-like flower.

8

Liliaceae
(Lily family)

Lilium philadelphicum   Western red lily (Plate 13)

The red lily which may be found in wet valleys and around lakes in the sandhills belongs to this species. In our plants the leaves are narrower and tend not to be arranged strictly in whorls as is characteristic of plants growing in the eastern United States and thus they have been assigned to variety *andinum.* The shoot arises from a bulb about an inch in diameter. The reddish orange blossoms radiate from the tip of the stem and usually number from one to three per plant. The tepals have lanceolate blades supported on slender stalks. This produces a flower with an open cup in contrast to the closed tubes of some lilies. The blades of the tepals are orange, fading to yellow near the broad base, and this area usually is spotted with darker colors. The stalk or claw of each tepal is provided with a nectar groove. Flowers mature during June and July.

Liliaceae
(Lily family)

Polygonatum biflorum   Solomon's seal (Plate 14)

These plants are common in moist woods and along streams throughout the state.

The gracefully arched stems from two to six feet long arise from a horizontal rootstock. They bear many sessile, alternate, ovate to oblong leaves two to six inches long, displayed in rows along the sides of the stem. The stems and leaves are smooth and hairless.

The specific name implies that flowers occur in pairs, but actually in our large plants the number of flowers in an umbel from the axil of a leaf varies from two to ten. Most of the plants of this species in Nebraska are apparently tetraploid (having twice the normal number of chromosomes in the nucleus of each cell) and have formerly been referred to the species *P. commutatum.* Most often there are three to five greenish-white flowers in a cluster hanging beneath the stem at many of the nodes. The tepals are united into a six-lobed tube which may be almost an inch long. Six

9

stamens are attached to the tube, and the three-celled ovary, after pollination and fertilization, develops into a dark blue berry. Most of our plants flower during May and June, and the berries persist until the plant withers in late summer or fall.

Liliaceae
(Lily family)

Smilacina stellata   False Solomon's seal (Plate 15)

Throughout most of northern North America, as well as this state, this plant may be found growing in woods or meadows, especially in sandy soil. It bears flowers in a terminal inflorescence called a raceme. Each small, white, lily-like flower may not exceed one-fourth inch in width but seems to be so perfectly formed and regular in appearance that the species name meaning star-like has been given to this plant. Flowering continues from May through July depending upon the habitat and the particular season. The berries are green at first and then become striped with six black bands. Later the fruit becomes uniformly dark red and is usually less than one-half inch in diameter.

The vegetative aspect of this plant is reminiscent of Solomon's seal because of its two rows of sessile clasping leaves and thus the common name, but the terminal versus the axillary flowers should distinguish the two genera with a second glance.

Another species, *S. racemosa,* with a branching inflorescence and red berries, is almost as common in Nebraska and extends farther south in the United States.

Liliaceae
(Lily family)

Trillium nivale   Dwarf white trillium (Plate 16)

This plant has also been called the snow trillium in deference to the specific name which means growing near the snow line. Another common name, early wake-robin, refers to its early appearance in March before many other plants have started their seasonal activities. The trilliums are rare in the state and only occasionally is it possible to find any number of them growing in one location.

10

The three petioled leaves which are usually not mottled, the low growth (three to six inches), the peduncled flower, the three greenish sepals, the three white petals, and the three-lobed ovary which is not winged, serve to distinguish this plant from other early spring plants and from other species of trillium.

T. grandiflorum is a larger plant also having stalked white flowers and often found in spring flower gardens, but it is not native in Nebraska.

T. recurvatum is small but has sessile flowers, dark red petals, and mottled leaves. It may also be found in cultivation but is not native here.

Liliaceae
(Lily family)

Yucca glauca   Yucca (Plate 17)

Although this plant may be known by many names such as bear grass, soapweed, spanish bayonet, and grass cactus, the simple name yucca seems quite sufficient. At home in dry, well-drained soil, it may be found in most parts of the state but excels in its stately, picturesque habit in the spectacular pine ridge country of northwestern Nebraska where it flowers during June.

The success of seed production in this plant is dependent upon a tiny, silvery moth which inhabits the flowers. The moth gathers balls of pollen and places them on the stigma. This ensures pollination but the moth deposits eggs within the ovary of the flower. A whole row or two of ovules or young seeds in the ovary are consumed by the larvae of the moth, but a sufficient number are left to produce many seeds in the mature capsule. It is stimulating to speculate on how such an association got established and how it has persisted to the present time.

Liliaceae
(Lily family)

Zigadenus nuttallii   Death-camas (Plate 18)

As the common name implies, this plant is poisonous. It grows in the dry, open areas of the prairies of western Nebraska and blossoms in May or June. The flowers are borne in a single, stout ra-

11

ceme (it may occasionally be a panicle) with the individual flowers on stalks less than an inch long. Most frequently a plant produces but a single inflorescence during a year's growth. This arises from a group of basal leathery, often recurved, leaves which may attain a length of one foot or more.

The yellowish white tepals are ovate to oval, bear obovate glands, and are notched at their ends. The stamens are separate from one another. The capsule produces numerous seeds.

Amaryllidaceae
(Amaryllis family)

Hypoxis hirsuta   Yellow star-grass (Plate 19)

Of course this small plant is not really a grass; rather it belongs to the amaryllis family whose members resemble those of the lily family except for the technical difference that the ovary is inferior rather than superior. This means that the floral tube is attached to and is a part of the ovary wall instead of being free or separated from it. The perianth and the stamens separate at a level above that of the ovary.

The bright yellow segments of the flower radiate in a symmetrical pattern with a stamen adjacent to the base of each sepal and each petal. When sepals and petals look as much alike as they do in *Hypoxis* the technical manuals usually call them all tepals. Although plants may be found blossoming through most of the growing season, the main flush may be observed in May and June in meadows throughout most of the state.

This is not a conspicuous plant since often only one or two flowers on a flower stalk are open at a time, and these usually are an inch or less in diameter. The common name has probably come from the regular yellow flowers and the long narrow leaves which at first glance may resemble grass leaves. (In the picture the broad leaf near the flower does not belong to this plant.)

12

Iridaceae
(Iris family)
Sisyrinchium angustifolium   Blue-eyed grass (Plate 20)

This flower was photographed in Chadron State Park in June. The deep blue color characterizes this species in contrast to the lighter blue to white flowers of *S. campestre,* found commonly in eastern Nebraska. This plant is a member of the iris family, but its flower parts are not as highly modified as those of the genus *Iris.*

The flowers arise from a pair of spathes which often terminate the scape. The flat scape is often said to be winged. In the field, a hand lens greatly enhances the beauty to be seen in these flowers with the blue tepals surrounding the bright yellow center. The filaments form a tube at the end of which the anthers are attached. The slender stigmas may be seen to protrude from between the anthers.

Orchidaceae
(Orchid family)
Corallorhiza maculata   Coral-root (Plate 21)

In this species, the plants are about a foot tall with tiny bi-colored flowers arising from the reddish or purplish scape forming a slender raceme. The rootstock is toothed and coral-like, hence the scientific name. As is the case in many orchids, this plant is a root parasite. A few scales surround the base of the scape but there are no prominent green leaves.

In August or September in cool, shaded ravines along the Missouri River this plant can be found in blossom. However, it has also been collected from the canyons of the northern and western counties of the state. The picture, Plate 21, was taken near the Missouri River on October 1.

Orchidaceae
(Orchid family)
Cypripedium calceolus   Yellow lady-slipper (Plate 22)

The photograph of this plant was taken in Indian Cave State Park, located in the northeastern corner of Richardson County,

13

Nebraska. Naturalists have seen this plant at various points along the Missouri River bluffs within the state in cool, damp woods. The yellow lady-slipper which some years ago was fairly common is now a rare species in this state. These plants most commonly have a single flower on stems up to two feet tall. There may be several sheathing elliptical leaves three to eight inches long below the terminal flower. The leaf-like bract which subtends the flower stands erect. There are two apparent spreading, brown sepals, about two inches long. The upper one is ovate, undulate, and pointed. The lower one is narrower and is composed of two united sepals. The lateral petals are narrow, greenish to brown, spreading and about as long as the sepals. The median petal is designated as a lip, is shorter, and is yellow with purplish markings. These orchids flower in May or June.

Orchids are rare within our state and should not be disturbed. More often than not, they die as a result of attempted transplanting. If you enjoy orchids take your hand lens along and visit your favorite haunts but avoid trampling surrounding vegetation as much as possible.

Orchidaceae
(Orchid family)
Orchis spectabilis   Showy orchis (Plate 23)

Along the wooded bluffs on the Missouri River near Peru and Plattsmouth this orchid has been collected. Two large, broad, persistent, basal green leaves are characteristic of this plant. The flowering stalk bears flowers as early as May or June and may remain erect until after frost. In the photograph one may even see the old seed pod from the previous year's growth. The delicate lilac or purplish color of the sepals and lateral petals, along with the white lip, provides a rare beauty which enchants the true flower lover when he first comes upon this plant.

Orchidaceae
(Orchid family)
Spiranthes cernua   Nodding ladies' tresses (Plate 24)

In late September, a month or so after the last hay crop has

Plate 25 Wood nettle
Laportea canadensis

Plate 24 Nodding ladies' tresses
Spiranthes cernua

Plate 26 Bastard toad-flax
Comandra umbellata

Plate 27 Umbrella plant
Eriogonum annuum

Plate 28 Swamp smartweed
Polygonum coccineum

Plate 30   Winged dock
Rumex venosus

Plate 29   Climbing false buckwheat
Polygonum scandens

Plate 31   Winged-pigweed
Cycloloma atriplicifolium

Plate 32   Slender froelichia
Froelichia gracilis

Plate 33   Heart's-delight
Abronia fragrans

Plate 34   Hairy four-o'clock
Mirabilis hirsuta

Plate 35   Wild four-o'clock
Mirabilis nyctaginea

Plate 36   Pokeweed
Phytolacca americana

Plate 37   Spring beauty
Claytonia virginica

Plate 38   Sandwort
Arenaria hookeri

Plate 39   Deptford pink
Dianthus armeria

Plate 41   Catchfly
Silene noctiflora

Plate 40   Bouncing Bet
Saponaria officinalis

Plate 42   Chickweed
Stellaria media

Plate 43   Spatterdock
Nuphar advena

Plate 44   White water-lily
Nymphaea tuberosa

Plate 45   Meadow anemone
Anemone canadensis

been removed from the meadows of Pierce County, one may look across vast acres of land dotted with little white flags. These are racemes of white orchid blossoms of this interesting plant. The clockwise, counterclockwise, and straight rows of flowers on the erect spikes can afford a flower lover hours of quandary and delight. The fragrant flowers less than one-half inch long are in two or three spirals. If a technical manual and a hand lens are available many minute details may be observed. These plants don't need the attention or protection provided by your private garden. They have been here many years; let them stay a while longer!

Dicotyledons

Urticaceae
(Nettle family)

Laportea canadensis   Wood nettle (Plate 25)

The ground in wooded areas may be densely covered with a growth of wood nettles. The leaves are broad and ovate and are alternately arranged on the stem. They are supported on long petioles and are covered with fine stinging hairs. The leaf margins and tip are sharply toothed. Unisexual flowers may be separated or mixed on the same plant. The clusters of staminate flowers, arising from the axils of the lower leaves, are seldom as long as the petiole of the leaf. The pistillate flowers from the upper leaf axils are in branched flower clusters longer than the petioles. The small inconspicuous flowers are in bloom during July and August.

Santalaceae
(Sandalwood family)

Comandra umbellata   Bastard toad-flax (Plate 26)

One hardly suspects such a prim, stiffly erect and wiry branched herb of being a parasite, yet it apparently survives in our prairies

15

by drawing its nourishment from the roots of other plants. The inch-long bright green, oblong leaves have entire margins and are sessile. The white flower clusters, occurring in May and June, terminate the branched stems. Flowers lack petals but have oblong sepals which unite below to form a floral tube. The stems arise from an underground rootstock.

Polygonaceae
(Buckwheat family)
Eriogonum annuum   Umbrella plant (Plate 27)

As a very common plant of the sandhills, the umbrella plant exhibits two striking features. It stands erect and is white, yet to the touch it is soft and pliable. Although manuals list several species of this genus for our area, this species is the most common. The plants often are widely separated, not growing in groups or clusters. The whiteness of this species results from the covering of white woolly hairs which clothe all parts above ground. Plants are one to three feet tall and bear tiny flowers which are arranged in a flattish cluster at the top of the plant, giving the common name to this plant. The small white flowers are six-parted and the fruits three-angled with a single seed. These plants are conspicuous from July on through the growing season.

Polygonaceae
(Buckwheat family)
Polygonum coccineum   Swamp smartweed (Plate 28)

Commonly seen from midsummer to fall as one drives along the low areas of the Platte River are the masses of these pink flowers which beautify otherwise swampy or unsightly borrow pits or roadside ditches. This smartweed is also to be found in several of our reservoirs as well as around sandhill lakes and farm ponds. The erect flowering shoots distinguish several closely related species of this genus. *Polygonum pensylvanicum,* growing in somewhat drier areas, is a common weed in the state. Plants with two types of flowers (long styles and short stamens or short styles and long stamens) probably belong to *Polygonum longistylum.*

16

Polygonaceae
(Buckwheat family)
Polygonum scandens   Climbing false buckwheat (Plate 29)

This scrambling vine inhabits thickets and covers vegetation in old roadside ditches. Like most members of this genus, it has such small flowers as to be hardly worthy of inclusion in a wild flower book, but the flowers all together are often showy. This species looks like a bindweed because of the general appearance of the leaves. However, the edges of the leaf are convex rather than concave as in the true bindweeds. Leaf bases vary from heart-shaped to sagittate and leaf tips are long and narrow.

The white flowers are produced in axillary clusters near the ends of the stem during August and September. Jet black, shiny fruits soon develop within the winged calyx. The similarity of these fruits to those of domestic buckwheat in shape and taste gives rise to the common name.

Polygonaceae
(Buckwheat family)
Rumex venosus   Winged dock (Plate 30)

Of the many weedy species of this genus, winged dock is one of the most spectacular in the sandhill region of the state. Again, in this genus the flowers are not conspicuous, but the brilliant red color comes from the wings of the fruit. This species has also been called wild begonia because of the three-angled ovary and fruit. As a low-growing perennial it arises at intervals from a woody rootstock. It is found growing in the western part of the state in waste places or cultivated fields and in prairie and pasture land. One may find it in flower or fruit throughout most of the summer.

Chenopodiaceae
(Goosefoot family)
Cycloloma atriplicifolium   Winged-pigweed (Plate 31)

One of the tumbleweeds of our state is this symmetrical low-growing member of the Chenopodiaceae. This is another family of

17

plants having inconspicuous flowers. This species is often found in low places in the sandhills of the state. Often these grow in waste places either singly or with a few clustered together. The most conspicuous color aspect of this plant is its change from green in early summer to purplish or reddish when mature in late summer.

Amaranthaceae
(Amaranth family)

Froelichia gracilis   Slender froelichia (Plate 32)

As indicated by the common name, this plant is of slender habit and is usually erect except for the nodding inflorescence at the summit of the shoot. The flowers are not conspicuous but the whole plant is soft with a clothing of white, woolly hairs. This covering is probably responsible for another common name, snake cotton, which is often used for this genus. The upright stem arises from a cluster of basal leaves and sometimes, though rarely, attains a height of two feet. The linear, oblong leaves occurring in pairs at the nodes of the stem have entire margins. They are from one-half to two inches long.

This is another family in which the flowers lack petals, but the calyx is tubular and also woolly. The five stamens have their filaments united into a tube. The five crests of the fruiting calyx are interrupted, forming distinct spines. These plants flower from June to September.

Nyctaginaceae
(Four-o'clock family)

Abronia fragrans   Heart's-delight (Plate 33)

These sprawling plants grow in dry soil in the sandhills and on the plains of the western part of Nebraska, where they are common. They flower from June to August. Although they are not related to verbenas, they have been called sand verbena or sweet sand verbena. Another common name used by ranchers of the state is prairie snowball.

The stems are spreading and are covered with sticky hairs. The leaves are in pairs but are often of unequal size. They are simple, entire, and bear no stipules.

18

The sweet-smelling white flowers are arranged in a head which is subtended by distinct bracts. Each regular flower is about an inch long, and the simple perianth appears corolla-like although the showy parts are technically considered sepals. Each flower has four or five stamens and a pistil that matures into a winged fruit almost three-eighths of an inch long.

Nyctaginaceae
(Four-o'clock family)

Mirabilis hirsuta   Hairy four-o'clock (Plate 34)

This four-o'clock is common in the dry valleys of the sandhills and on dry hills over the western part of the state. The erect stems grow to over two feet in height. They are hirsute and become glandular-hairy in the inflorescence. The lanceolate leaves are sessile except for the lowest ones which are supported on short petioles. They occur in pairs and are equal in size.

The involucres consist of five united bracts forming a broad and open cup subtending three to five flowers. In fruit, the involucre enlarges and becomes veiny. The flowers appear in July and August. As is characteristic of this family, there is no corolla but the calyx has the appearance of a corolla, and the involucre often appears to be a calyx. The calyx tube is very short and the limb is rotate, pink to purple in color, and hairy on the outside. There are often five stamens in a flower. The fruit, which matures from the ovary, is indehiscent, club-shaped, and angled. The angles are smooth, but otherwise the fruit is hairy.

Nyctaginaceae
(Four-o'clock family)

Mirabilis nyctaginea   Wild four-o'clock (Plate 35)

Considered a weed by many people, this plant is a perennial possessing a fleshy tap root and reproducing by seeds. The stout, dichotomously branched, smooth stems bear heart-shaped leaves. The flowers are technically somewhat different from what might be expected. Green parts in the outer whorl are bracts, and the small red parts of the flower are sepals. Petals are not present in these flowers. The fruits are hairy and often club-shaped with five ribs.

Phytolaccaceae
(Pokeweed family)

Phytolacca americana   Pokeweed (Plate 36)

Pokeweeds, which often grow to heights of more than six feet, have rank, fleshy, upright stems bearing racemes of flowers in July or August and dark-juiced berries about one month later. The leaves are smooth, large, and veiny. Flowers are small and not as conspicuous as the berries. The sepals are the colorful parts of the flower, being white or tinged with pink. Petals are absent. The roots are large and are reported to be poisonous.

Portulacaceae
(Purslane family)

Claytonia virginica   Spring beauty (Plate 37)

Depending upon exposure to sunlight, the flowers of the spring beauty vary from pink to white. The veins of the five oval petals always appear to be a darker pink. The flowers occur in a loose raceme at the summit of a comparatively weak stem bearing a single pair of opposite leaves which are long, narrow, and nearly sessile. The stems and basal leaves arise from a small corm as early as March and frequently flowering has finished before May. In Nebraska they are found only in damp woods along the Missouri River.

Caryophyllaceae
(Pink family)

Arenaria hookeri   Sandwort (Plate 38)

This plant, growing in rather low, dense mats from one to four inches high is frequently found in northwestern Nebraska. Visitors to Chadron State Park before or near the Fourth of July usually have a good chance of seeing it along the rocky footpaths among the pines, at the higher elevations.

Seen from a distance sandwort is easily confused with *Phlox hoodii* or *P. andicola*, but the separate petals of this species should quickly be distinguished from the tubular corolla of *Phlox* on

20

close examination. Between each of the five white petals may be seen the sharp points of the five green sepals which are about the same length as the petals. The ten stamens, having long filaments, are also quite different from those found in *Phlox*.

Caryophyllaceae
(Pink family)

Dianthus armeria   Deptford pink (Plate 39)

Deptford pinks are native to Europe but they have become naturalized in North America as far west as Ontario and Oklahoma. Growing as annuals or biennials in pastures, prairies, roadsides, and open woodland, they flower at almost any time during the growing season, from May to October. The middle of June finds them at the peak of their floral display in prairie areas in eastern Nebraska. By July most of the plants bear fruit with only an occasional flower.

These are delightful, slender plants brightening their area with small, vivid flowers among an otherwise green background of developing grasses. They do not seem to be strong competitors since they are more prolific in a dry or overcropped prairie than in dense sod of tall grasses. Their leaves are linear and the stems pilose.

The inflorescence consists of a few to several flowers closely crowded in cymose clusters. Characteristically, only one flower of a cluster is open at a time, as is well illustrated by Plate 39. There are several pointed bracts, usually in pairs, below each flower.

Caryophyllaceae
(Pink family)

Saponaria officinalis   Bouncing Bet (Plate 40)

Although this plant is not native it has become naturalized from Europe. It grows well in Nebraska, blooming in July and August, and it commonly appears in roadside areas and along railroad embankments. It spreads by an underground rootstock and produces stems about two feet tall with opposite leaves which have several

21

longitudinal ribs. The flowers are abundant at the top of the stem, each being about one inch wide. The flowers are pink or white, often intermixed on the same plant. The calyx surrounds the capsule which produces numerous seeds. Bouncing Bet was a favorite in the flower gardens of the early settlers of this area.

Caryophyllaceae
(Pink family)

Silene noctiflora   Catchfly (Plate 41)

This plant grows as a medium-sized, conspicuously hairy weed of waste places and clover fields or meadows. It has opposite leaves and white flowers which have deeply notched petals and a bladdery, ribbed calyx. The blooms open in the evening during the months of July, August, and September. These plants often grow in thick patches, crowding out other plants. Flowers are unisexual and the plants are usually less than three feet tall. The pistillate flowers have three styles. The capsules usually produce an abundance of seeds. Often confused with this plant is *Lychnis alba* which is dioecious and whose pistillate flowers have five styles.

Caryophyllaceae
(Pink family)

Stellaria media   Chickweed (Plate 42)

Several plants are known by the common name chickweed. The genus pictured on Plate 42 is *Stellaria*. In contrast to other chickweeds the leaves of this one are nearly glabrous except for the ciliate margins. These weeds thrive in moist or sheltered conditions through most of the seasons of the year, though they are technically annuals. They usually have five white petals which may be so deeply divided as to appear as ten petals. The number of styles is usually three, and the capsules open at maturity by splitting to the base.

Another genus, *Cerastium,* which is similar, differs in that the capsule splits open only at the summit, the leaves are viscid pubescent, and the plant is perennial.

22

Nymphaeaceae
(Water-lily family)
Nuphar advena   Yellow pond lily, spatterdock (Plate 43)

In a revision of this genus by Beal (1956) this plant becomes a variety of *Nuphar luteum*. The emersed leaves borne on stout petioles have triangular sinuses separating the basal lobes of the generally circular leaf blades. The petioles are circular in cross section rather than oblong as in other species. The leaf blade has a more definite midrib than other water lilies. Its most common habitat is our sandhill lakes or ponds. Flowering may occur any time during the summer.

Emersed like the leaves, the flowers are held on stout peduncles which arise, as do the petioles, from a heavy rhizome. There are six sepals surrounding the other floral parts. Sepals which are exposed are green on the outside and yellow around the edge and on the inside. They are strongly curved, forming a globose flower bud. The petals are numerous and shorter than the stamens and usually remain unnoticed in the flower. The stamens are also numerous with flat filaments which are shorter than the anthers. The prominent, compound ovary in the center of the flower is prolonged into a thick style and a broad disk-like stigma with about seventeen rays (as shown in the picture on Plate 43).

Nymphaeaceae
(Water-lily family)
Nymphaea tuberosa   White water-lily (Plate 44)

The white water-lily grows in quiet waters like those near Loup City and the Two Rivers Recreation Area. The flowers have four greenish sepals and numerous blunt-tipped white petals. Many stamens grading from one type to another are found within each flower. The inner stamens have linear filaments and long anthers while the outer filaments are petal-like and have shorter anthers. The ovary is compound and is surmounted by several radiating stigmas. The flowers lack an odor. Flowering may occur from June to October.

The leaves as well as the flower buds may emerge from the water but with age they usually float on the surface. The leaf blades

23

have a deep, narrow sinus and are the same color on the lower surface as on the upper. The petioles are green but near the blade they are striped with brown.

Ranunculaceae
(Crowfoot family)
Anemone canadensis   Meadow anemone (Plate 45)

This is a white-flowered plant having a few white sepals representing the showy part of the flower. The pistils are separate but are aggregated into a globose head. Each of the pistils has a stout style. This is one of the more attractive anemones because of its profusion in meadows and the large size of the white sepals. The flowers are supported at varying distances above the sessile, involucral leaves. Blossoms occur in June and even into July depending on the season and the abundance of moisture.

Ranunculaceae
(Crowfoot family)
Anemone caroliniana   Prairie anemone (Plate 46)

This is one of the early-blooming anemones (April-May) which may have white blooms but frequently has colored flowers, as pictured here. It is one of the small anemones often seen with only rosettes of basal leaves which are usually divided into three main segments (ternately divided). These in turn may be more or less incised. The involucral leaves lack petioles and are thus said to be sessile on the flowering shoot where they occur below the middle of the stem. Numerous colored flower parts, which in this genus are technically sepals instead of petals, are distinctive for this species. There is usually but one flower per plant.

Ranunculaceae
(Crowfoot family)
Anemone cylindrica   Thimble weed (Plate 47)

The common name of this plant, as well as the specific name, is derived from the long and cylindrical shape of the flower head. In

24

spite of its common name, it is not actually very weedy. Frequently found over the entire state, it is one of the later-blooming species, often not in flower until July. The sepals are smaller than in the previous species and they often tend to be more greenish white than white. These plants are of medium height with the flowers six to twelve inches above the petioled, involucral leaves. As pictured, there may be several flowering stalks arising from the upper nodes of the stem. A fairly characteristic structure of the leaf is that the segments have straight, more or less parallel edges toward their bases.

Anemone virginiana resembles the above species, except that the leaf segments are wider, more tapered, and incised to below the middle.

Ranunculaceae
(Crowfoot family)

Anemone patens   Pasque flower (Plate 48)

The pasque flower is one of our earliest-blooming plants of the prairie in northern and western parts of the state where it comes into flower in March or April. The distinctive appearance of this anemone is shown in the accompanying picture taken from ground level which indicates that its stature is short and that the whole plant is covered with silky hairs. Its large showy flowers have five to seven purplish or lavender sepals which open to two or three inches in diameter. Fluffy, hairy seed heads follow the flowers and persist until June or July. This is the state flower of South Dakota.

Ranunculaceae
(Crowfoot family)

Aquilegia canadensis   Columbine (Plate 49)

Flowers are crimson to pink with yellowish areas, and the blooms have five spurs, each of which terminates in a knob-like nectar gland. These are delicate plants of open woods and rocky slopes. The leaves are twice ternately divided with each ultimate segment being somewhat wedge-shaped. They bloom from April to June, but you may be disappointed if you look for them after the middle of May. They are not uncommon along the Missouri River.

The blue-flowered columbine, *A. caerulea*, is the state flower of Colorado.

25

Ranunculaceae
(Crowfoot family)

Clematis virginiana   Clematis or virgin's-bower (Plate 50)

Virgin's-bower is a common vine found growing in low, moist ground along streams or around ponds and hedgerows. It supports itself by twining petioles as it scrambles over shrubs or even into trees. The clusters of white flowers and leaves of three segments are characteristic of this species. It is in flower during July, and by August the whole vine may be almost hidden by a profusion of feathery tailed fruits which have developed from the numerous pistils of each flower.

Ranunculaceae
(Crowfoot family)

Delphinium virescens   Prairie larkspur (Plate 51)

This white to pale blue larkspur may be found in the full flush of its blooming period during June. The single, erect stalks bearing many flowers arise before the grasses have put forth their first rapid growth of leaves. Often by the time the last flowers have faded, however, the grasses have surrounded and nearly hidden the larkspurs which earlier stood as separate, short sentinels over many acres of Nebraska prairies. The intricacies of the flower structure may fill you with great delight as you pick a single flower and examine its parts to find the double nature of the spur. This may arouse questions about adaptations on which many a sunny hour may be spent.

Leaves are alternate and are palmately divided. The flowers are in a terminal raceme and in each one the posterior sepal is prolonged into a spur. The sepals form the externally visible part of each flower. Three or more follicles which are joined by their bases and are about four times as long as broad may be produced from a single flower.

26

Ranunculaceae
(Crowfoot family)
Ranunculus circinatus   White water crowfoot (Plate 52)

As indicated by the common name, this is a white-blossomed aquatic plant. The leaves are dissected into many small divisions and are entirely submerged. The flowers arise singly as indicated in the accompanying illustration and emerge from the water. They exhibit five shiny, white petals, many yellow stamens, and many pistils in a globular head. This species has leaves which are sessile and whose parts do not mat together when lifted from the water. These leaf characteristics distinguish this species from *R. aquatilis* whose leaves are petioled and whose leaf segments do mat together when removed from water.

Ranunculaceae
(Crowfoot family)
Ranunculus sceleratus   Cursed crowfoot (Plate 53)

This plant is to be found in mud flats, sloughs, and swampy depressions and around oxbow lakes where its vegetative parts are often submerged during certain seasons. Having small yellow petals and a tall receptacle on which the pistils are attached, the flowers are not spectacular in the least; but there are many flowers on each stalk. When looking at rank vegetative growth, one may almost see a resemblance to celery stalks. However, the specific name means acrid or biting and the plant juice will cause blisters on the skin and may poison cattle if eaten. In any case, dairy cows grazing these plants will have very bitter milk.

Berberidaceae
(Barberry family)
Podophyllum peltatum   May apple (Plate 54)

In the southeastern part of Nebraska, in deep woods, one finds the heavy leaves of the May apple making their appearance in late April or May. Most of the stalks support but a single leaf, but where a branch occurs and two leaves are produced one may ex-

27

pect to find a flower on a short stalk arising between the bases of the two leaves. The flowers are white with six or more petals and twice as many stamens as petals. The stigma is attached directly to the top of the ovary, omitting the style possessed by most flowers. The ovary develops into a fruit which is edible, although other parts of the plant are supposed to be somewhat poisonous.

Papaveraceae
(Poppy family)

Argemone polyanthemos   Prickly poppy (Plate 55)

A common weed of disturbed areas in the central and western part of the state, this is not easily mistaken for any other plant because of its distinctly poppy-like flower with four showy white petals and numerous yellow stamens. It has spiny stems, leaves, and sepals which account for the common name. Upon removing a leaf or otherwise cutting into this plant one may be surprised by the yellow color of the latex which readily flows from the wounds. The flowers are large and showy and thus may be recognized from a car window even at moderate speeds. The most common month for flowering is either June or July depending upon the season and the location in the state, but earliest and latest flowering extends this period from May to August.

Papaveraceae
(Poppy family)

Sanguinaria canadensis   Bloodroot (Plate 56)

Bloodroot is a rare plant in Nebraska, but it is one of the delicate, early flowers of spring. The common name is derived from the colored juice which exudes from broken underground stems (rhizomes) when a plant is pulled up. A single leaf emerges from the ground in an erect position. It is loosely rolled around the flower stalk and often shows prominent veins on the under surface. After the flowering period, the leaves unroll, flatten out, and soon cover the small capsules produced by the flower. When mature, the leaves are more or less circular in outline and often have several lobes around the margin.

28

Typically the flower stalk bears a single flower an inch or more in diameter with eight petals of unequal size, many stamens, and a single pistil terminated by a two-lobed, capitate stigma. At maturity the capsule, derived from the ovary, is over an inch long. Flowering occurs in April or May. These plants are available commercially and grow well in a flower garden with a minimum of care.

Fumariaceae
(Fumitory family)

Corydalis aurea   Golden corydalis (Plate 57)

These weedy little early-flowering plants are among the first to produce flowers in the spring. In March or April these herbs take advantage of warm days, and a cluster of greenish yellow flowers is differentiated at the tip of a short stalk. The whole plant is often only four to twelve inches high. The finely incised leaves are characteristic of all species of this family. The leaves and stems are usually glaucous. These plants are more common in the eastern half of the state, inhabiting disturbed areas at the edges of plowed fields and along roadsides.

These flowers have two sepals, two outer petals enclosing two inner ones, six stamens, and a slender pistil which matures into a several-seeded pod. In this genus the upper petal forms a spur which makes the flower appear to be attached at about the middle. The outer pair of petals are merely keeled on the back and do not form flattish projections called crests which occur in *C. micrantha.*

Fumariaceae
(Fumitory family)

Dicentra cucullaria   Dutchman's-breeches (Plate 58)

On well-drained hillsides in rich wooded areas these small early plants are to be found in full bloom by the middle of April. The bluish green, highly dissected leaves form their characteristic patterns over the dried tree leaves from the previous year. From a cluster of small tubers rises an erect scape bearing several flowers which are bilaterally symmetrical and hang down from slender branches called pedicels. At maturity the main flower stalk is

29

often curved or bent with the flowers it supports. The two outer petals form the spurs which make the flower resemble an inverted pair of breeches because of their spreading tips.

Capparidaceae
(Caper family)

Cleome serrulata   Rocky Mountain bee plant, pink cleome, beeflower, Indian pink (Plate 59)

Although pink and white petals are common in the same plant and even in the same flower, and though the flower clusters may appear from white to purple, pink is by far the most common aspect of this species. The handsome flowers are borne in elongating terminal flower clusters called racemes. Simple bracts subtend the flowers which are supported on elongate flower stalks. Each flower has six stamens whose long slender filaments greatly exceed the length of the four petals. The fruit pods from the older flowers of a cluster are usually quite conspicuous because of their length. They bear many roughened bean-shaped seeds upon which the next year's plants are dependent, for these plants are annuals, and are often found in large stands in the prairies. Flowers occur in July and August.

Another species, *C. lutea,* with yellow flowers has been found in the state but not frequently. It has five leaflets per leaf instead of the three leaflets which are generally found in pink cleome.

Cruciferae
(Mustard family)

Berteroa incana   Hoary alyssum (Plate 60)

As an introduced plant *Berteroa* has become established in various places in the state. The picture on Plate 60, taken in the northern part of Cherry County, south of Cody between the Snake and Niobrara rivers, represents a small portion of quite a large cluster of plants. For an annual, the stems of this plant are more woody than usual. The long, flowering portion of erect stems bears numerous small white flowers having four petals, each with a deep

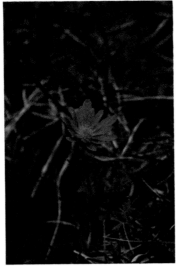

Plate 46   Prairie anemone
Anemone caroliniana

Plate 47   Thimble weed
Anemone cylindrica

Plate 48   Pasque flower
Anemone patens

Plate 49   Columbine
Aquilegia canadensis

Plate 50   Virgin's-bower
Clematis virginiana

Plate 51   Prairie larkspur
Delphinium virescens

Plate 52   White water crowfoot
Ranunculus circinatus

Plate 53   Cursed crowfoot
Ranunculus sceleratus

Plate 54   May apple
Podophyllum peltatum

Plate 55   Prickly poppy
Argemone polyanthemos

Plate 56   Bloodroot
Sanguinaria canadensis

Plate 57   Golden corydalis
Corydalis aurea

te 58   Dutchman's-breeches
centra cucullaria

Plate 59   Rocky mountain bee plant,
Cleome serrulata

Plate 61   Indian mustard
Brassica juncea

Plate 60   Hoary alyssum
Berteroa incana

Plate 62   Toothwort
Dentaria laciniata

Plate 63   Western wallflower
Erysimum asperum

Plate 64   Dame's rocket
Hesperis matronalis

Plate 65   Silvery bladderpod
Lesquerella ludoviciana

Plate 66   Water cress
Rorippa nasturtium-aquaticum

Plate 67   Tumble mustard
Sisymbrium loeselii

notch in it. Below the flowers numerous fruits occur. These are elliptical or oval in outline and are usually rounded in cross section, although the seeds within the pod are generally flattened. This plant differs from the bladderpod in its white flowers, notched petals, and a persistent style which is shorter than the fruit. June until September are the months in which it blooms.

Cruciferae
(Mustard family)

Brassica juncea   Indian mustard (Plate 61)

Indian mustard is a glabrous annual that occasionally has a whitish waxy bloom on the surface of stems and leaves. The lower leaves of the plant are often quite large and this has given rise to the common name leaf mustard. The name brown mustard has also been applied to this species because of the reticulate brown seeds that are produced in which pods may be up to two inches long. When found in cultivated fields the brilliant yellow mass of flowers may make one forget for a few moments that these are weeds in the eyes of the farmer. This plant may bloom from July until October.

Cruciferae
(Mustard family)

Dentaria laciniata   Toothwort (Plate 62)

Toothwort is one of the spring flowers of our moist woodlands. Although basal leaves are produced, they are usually absent by the time the plant is in flower. By flowering time there are only two or three leaves on the stem and they appear five-parted and are irregularly toothed. The stalk bears several large flowers which may not immediately be recognized as belonging to this family. They are rarely open very wide and their color varies from white to a delicate light purple. In a late spring season one may find these in flower from the last of April into May. This plant is a perennial which spreads by the growth of a segmented rhizome.

31

Cruciferae
(Mustard family)
Erysimum asperum   Western wallflower (Plate 63)

The western wallflower is one of the most conspicuous crucifers in the western part of Nebraska. The large, bright yellow flowers shown in the picture are in a young inflorescence. Those of an older inflorescence have flowers considerably more diffuse and less noticeable. Flowering occurs during May and June with the production of slim seed pods four inches long in the lower part of the inflorescence while flowers are still opening at the top of the stem.

Another closely allied species is *E. repandum* in which the single flowers and the whole flower cluster are smaller than the above species. The petals are not as yellow and the seed pods seem to protrude stiffly from the zigzag stem almost at right angles.

Cruciferae
(Mustard family)
Hesperis matronalis   Dame's rocket (Plate 64)

The purple flowers of this crucifer make it one of the most colorful flowers of this family. It is found in eastern Nebraska but is not a native. It seems to have become naturalized in eastern United States from flower gardens. It is a perennial plant that produces stems which are branched above the base and terminate in a bright racemose flower cluster during June.

Upper leaves are smaller than the lower ones and all leaves have narrow pointed tips and wide bases. Leaves are either sessile or have short petioles and their margins have a few pointed teeth. With a good hand lens the branched nature of the hairs should be visible.

Flowers have four erect sepals, four purple petals with broad spreading blades and long claws, six stamens, and an elongate ovary with two erect stigma lobes. The fruit, which is a silique, is nearly cylindrical and bears a single row of seeds in each locule.

32

Cruciferae
(Mustard family)
Lesquerella ludoviciana　　　　　　　Silvery bladderpod (Plate 65)

This is another yellow-flowered member of the mustard family, and it commonly flowers during June. The fruits on this plant give the common name of bladderpod to this genus because of their inflated appearance. In this species the silvery, matted, branched hairs and the gracefully recurved stalks of the fruits are distinctive. The narrow, entire, alternate leaves may have wavy margins and are of varying lengths. The basal leaves may be as much as four inches long, but the upper leaves are progressively shorter. The whole plant is usually less than one foot tall when it first starts flowering, yet the stem continues to elongate as the fruits are produced. It is found frequently in the western part of the state.

Cruciferae
(Mustard family)
Rorippa nasturtium-aquaticum　　　　　　Water cress (Plate 66)

Water cress has been introduced from Europe and now grows in many parts of the state. It grows with the young stems submerged in water, and later these may arise above the water level. The leaves are compound with five to nine shiny, green, succulent segments or leaflets with toothed margins. As one might expect, there is considerable difference in the size, shape, and texture of submerged and aerial leaves and leaflets. Doubtless many people have collected the submerged parts without realizing that the plant has an aerial phase that bears flowers.

The flowers occur in clusters at the tips of the stems which are above water. The four white petals exceed the green sepals in length and the stamens are exserted from the tube of the corolla. The petal lobes are somewhat persistent in that the maturing pistil (fruit) extends out of the corolla tube while the petals are still white and expanded.

33

Cruciferae
(Mustard family)

Sisymbrium loeselii   Tumble mustard (Plate 67)

Tumble mustard is an introduced annual from Europe. It becomes a troublesome weed in certain areas of the state, taking over abandoned fields and waste places. The flowers are bright yellow but small, and they soon produce slender fruits on elongate peduncles. The plants are up to three feet tall with large basal leaves which are incised several times almost to the midrib. The leaves and stems are covered with stiff hairs, and near the base of the stem the hairs are recurved. In the eastern part of the state the largest flush of blossoms occurs during June, when a whole field may appear yellow.

Cruciferae
(Mustard family)

Thlaspi arvense   Pennycress (Plate 68)

This is one of the noxious weeds of Nebraska, but it is included here since it is conspicuous early in the growing season. Waste places appear white with the terminal clusters of flowers that appear in April or May. For part of the summer, the plants remain with their ripening fruits which are quite distinctive. The pod is circular or oval and flat with a notch at the upper end. It is broadly winged all around. The stem elongates as the fruits mature, greatly increasing the height of the plant. The leaves are simple, more or less toothed, and nearly sessile with small ear-like lobes (auricles) at the base.

Other white-flowered, weedy plants of this family include *Lepidium virginicum* (pepperweed), *Capsella bursa-pastoris* (shepherd's purse), and *Cardaria draba* (hoary cress). All of these have distinctive fruits which are well illustrated in *Nebraska Weeds,* a book published by the Weed and Seed Division of the State Department of Agriculture.

34

Crassulaceae
(Stonecrop family)

Penthorum sedoides   Virginia stonecrop (Plate 69)

Penthorum, sometimes also known as ditch stonecrop, is usually under two feet tall and has alternate leaves with small regularly spaced teeth. The leaves, narrowed at both ends, are about three-fourths of an inch wide and two to four inches long. The flowers occur in a forked terminal cluster usually of two or three branches instead of the four shown in the picture, with most of the flowers faced upward. They appear greenish because of the lack of petals and the presence of five sepals which are predominantly green. In the illustration, flowers with six carpels and twelve stamens may be seen, but most of them have five carpels and ten stamens. Though the styles and stigmas of each carpel are separate and radiately arranged, the bases of the ovaries seem to be united. This produces a flattened five-lobed capsule with as many horns. Flowering occurs mostly during July and August.

Rosaceae
(Rose family)

Fragaria virginiana   Wild strawberry (Plate 70)

Wild strawberries are often found growing in large colonies, having spread vegetatively by means of runners. Most of the blossoming occurs in May. They may be found on the prairie as it rolls off into a draw or in the shade of sumac as it grows on prairie slopes. There are few to several flowers on stalks of approximately equal length. These usually are lower than the leaflets which form a canopy for the flowers and later the ripening fruit. There are three leaflets per leaf and each has a serrate margin. On the mature fruits the achenes are set in pits on the red, enlarged mature receptacle which we eat as the fruit. In another species, *F. vesca,* more common in the western part of the state, the achenes are set on the surface of the receptacle.

The flowers have five narrow, green sepals which alternate with five broadly rounded, white petals. Numerous stamens surround the many pistils which are set on the receptacle.

Rosaceae
(Rose family)

Potentilla anserina   Silver-weed (Plate 71)

The common name is derived from the silvery appearance of the lower surface of the leaves which are covered by silky white hairs. The upper surfaces of the leaves are green. The whole plant may be sprawling on wet ground, at first without a stem; but later it puts out a stolon which sets new plants and produces peduncles, each bearing a single flower. Flowers have oval yellow petals which are longer than the ovate sepals. Twenty or more stamens are arranged in three series. This species has been placed by some authors in another genus, *Argentina,* because of the origin of the styles from the sides of the pistils rather than from the top as in other potentillas. This species is found in western Nebraska and scattered occasionally in other sections of the state. It flowers from May until August.

Rosaceae
(Rose family)

Potentilla arguta   Tall cinquefoil (Plate 72)

This is a hairy perennial that grows about three feet tall and has somewhat sticky leaves and stems. The pinnate leaves are compound, having from seven to eleven toothed leaflets along a midrib. Numerous flowers are set in close clusters at the upper nodes of the stem. Close examination of the flowers of this genus reveals that the green outer parts of the flower appear in two whorls. The outer parts are called bractlets and the inner whorl is composed of sepals. Normally in tall cinquefoil the white petals are longer than the sepals, and the sepals are longer than the bractlets. The many stamens and numerous pistils in the center give the flower a characteristic yellow spot.

There are several species of *Potentilla* that are similar in being erect plants with pinnate leaves, but they differ in other respects. Most species have flowers with yellow petals as well as yellow centers. These plants blossom in June and July and are most commonly found in dry woods and prairies.

36

Rosaceae
(Rose family)
Potentilla recta   Sulfur cinquefoil (Plate 73)

In Nebraska there are some species of this genus which have digitate (that is, palmately compound) leaves. The pictured species represents this group and illustrates one of the most rapidly increasing weeds in the state. It grows in meadows, pastures, hayfields, and waste places. It is a perennial plant and also produces quantities of achenes in each of its many flowers, by which it is rapidly spread. Although manuals describe this plant as having yellow petals there are many plants in which yellow is prevalent only at the throat. Petals are large and characteristically notched in the outer margin although this is not distinctive of this species alone. The flowers, with their numerous stamens and pistils and large petals, are conspicuous and attractive during their blooming season in June and July.

Members of this genus are low in forage value and are cropped very seldom by grazing animals.

Rosaceae
(Rose family)
Rosa suffulta   Prairie rose (Plate 74)

Prairie rose is closely allied or identical with the Arkansas rose and sunshine rose depending on how wide a species description you accept and what names are commonly used. *Nebraska Weeds* (1962) lists it as the wild rose *R. arkansana.* Fernald (1950) called it a variety of *R. arkansana,* while Gleason and Cronquist (1963) recognize it as a species as listed above.

This plant is the most common rose of ditches, roadsides, and prairies of eastern Nebraska. The above-ground parts are frozen back to ground level nearly every winter and new growth is produced each spring. The leaflets vary considerably in size depending, in part, on the amount of water available in the soil. The back sides of the leaves are covered by soft hairs.

Flowers are produced in small clusters with from one to three flowers of a cluster in blossom at the same time during the height of the season in June. The bud tips always appear a deeper pink

37

than the petals of the open flower which range from nearly white to a deep pink. The rose hips are usually bright red and remain colorful in late fall and winter.

Leguminosae
(Pea family)

Amorpha canescens   Lead plant (Plate 75)

Flowers occur in greatest abundance during July although the vegetative plant is prominent during June. Because of its strong stems it has been called prairie shoestrings. It is common in prairie areas throughout the state. The leaves are pinnately compound as is the case in many of the legumes considered in this book. The leaflets are of an odd number and small size. Their grayish appearance caused by hairs on the leaflets gives rise to the common name. Each flower is very small and reduced, having only one petal; but together they make a colorful display in the whole spike-like inflorescence. The stems become woody at the end of the season and new growth in the spring arises from woody rootstocks.

Leguminosae
(Pea family)

Astragalus canadensis   Milk vetch (Plate 76)

Although there are many species of *Astragalus* in Nebraska and the midwest in general, this species with its erect inflorescences of light yellow flowers is one of the most conspicuous as well as one of the most common. It is in bloom during July and August. Wild licorice has occasionally been confused with this plant, but most inflorescences will show a few lower flowers which have produced the rather smooth shiny rounded bean-like pods characteristic of this plant in contrast with the spiny bur-like fruit of the wild licorice. The leaflets are on the whole noticeably larger than those of licorice and considerably larger than those of other species of this genus commonly found in the state.

38

Leguminosae
(Pea family)

Astragalus ceramicus   Painted pod,
                           bird's-egg pea, rattleweed (Plate 77)

In the sandhills of Nebraska there are several species of plants resembling the one shown here. They differ in such technical details as the length of the stipe which supports the pod and the length of the pod itself. In more obvious features differences such as the color of the flowers, the coloring on the pod, and the number of leaflets are seen.

These low perennials may grow to a height of fifteen inches and persist even in active blowouts. The leaves are compound and bear a few filiform or linear leaflets but the upper leaves become reduced to a single linear rachis.

The flowers have the pea-like form common in this family and are white to light blue in color. The calyx forms a short tube ending in unequal sharp teeth. The pendulous pod is highly colored, ellipsoidal, and pointed at the tip. It has but one chamber within the paper walls, and usually several seeds develop within. The picture of this species was taken in June, but flowering continues from May to July.

Leguminosae
(Pea family)

Astragalus crassicarpus   Ground plum, buffalo pea (Plate 78)

In the early spring as the sunshine begins to warm the prairie sod one of the first legumes to be stimulated to flower is *Astragalus crassicarpus*. The vegetative parts are evident in March, and during April flowering occurs. This sequence may occur a month later in the northwestern part of the state. As the common names indicate, it is the large fruits, about an inch in diameter, rather than the flowers for which this species is noted. Instead of the normal elongated bean-pod type of fruit this plant bears a somewhat spherical fruit that resembles a plum in color and size. These lie on the ground since the vegetative plants are creeping or sprawling.

Leguminosae
(Pea family)

Astragalus mollissimus　　　　　　　　Woolly locoweed (Plate 79)

The common name, locoweed, refers to the poisonous effects upon cattle browsing these plants. Stems are decumbent and densely villous. Leaves are mostly basal and are pinnately compound. They bear twenty-three or more oval or obovate silky leaflets which are up to one inch long.

Bright purple flowers appear about the first of May, borne on short erect stalks. The keel, consisting of the two lower petals of the flower, encloses the stamens and pistil and is blunt at its tip. The pod produced from the pistil is cylindrical, under one inch long, and curves upward. These plants are found in the prairies of the central and western portions of the state.

Oxytropis lambertii, also called locoweed, is often confused with the species pictured. These plants lack stems and their pinnately compound leaves, bearing eleven or more lanceolate leaflets, arise directly from the soil. The flowers are dark bluish purple, have pointed keels, and appear as early as April in western Nebraska where they commonly occur.

Leguminosae
(Pea family)

Astragalus racemosus　　　　　　　Racemed milk vetch (Plate 80)

Characteristic features of this species, which show in the picture taken near Red Cloud, Nebraska, are the white to slightly lavender corolla, the distinctly drooping, three-angled pods, and the odd-pinnately compound leaves. Each leaf bears a terminal leaflet with seven to fifteen pairs of stalked leaflets below. An outstanding feature of this species is the distinctive appearance of the strictly erect, long-stalked racemes which undoubtedly are responsible for both the specific and the common names.

In central and western Nebraska where this plant is frequently found, it flowers during May and June. As shown in the picture, the upper flowers of the raceme are in blossom while the lower ones are forming fruits.

Leguminosae
(Pea family)

Astragalus adsurgens   Prairie milk vetch (Plate 81)

Flowering of this plant takes place in June and July. The picture was taken at the Cochran State Wayside in Dawes County south of Crawford. This is the same plant that some manuals have listed as *A. striatus.* The stems are relatively short and mostly ascending. Leaflets are elliptic and grayish because of their pubescence. Hairs are usually stiff, appressed, and mixed black and white. The flowers, as shown, are usually in short spikes, which are supported on peduncles that are shorter than the leaves beneath. Fruits after maturing are not much longer than the flowers and are covered with hairs similar to those found on the leaflets.

Leguminosae
(Pea family)

Baptisia leucantha   Large white wild indigo (Plate 82)

As indicated by the common name, the habit of this plant is much more erect and considerably taller than *Baptisia leucophaea.* Furthermore this plant has whitish flowers instead of yellow. Leaves of this plant are trifoliate, but the petioles are longer and stipules are not as large and are never mistaken as leaflets of the leaf. The leaflets are glabrous and the stems are glaucous.

The inflorescence is nearly erect and bears white flowers over a considerable length of the terminal flowering stalk. These plants are not too frequent and usually grow well separated from each other. In eastern Nebraska this plant is in flower in May and in fruit in June.

Leguminosae
(Pea family)

Baptisia leucophaea   Wild indigo (Plate 83)

The low widely branching nature of this perennial herb is usually distinctive enough to separate it from most of the other leguminous plants. The trifoliate leaves have short petioles and large leaf-like stipules which give the appearance of a leaf with five leaf-

41

lets instead of three leaflets and two stipules. The leaves are alternate on the stem, and their oblanceolate leaflets are covered with long soft hairs. Flowering occurs during April and May; however, some plants may produce another branch or two in June which bears flowers at the same time earlier flowers are maturing into fruits. At maturity fruits are mostly dark or blackish in appearance. Although early flowers may be a light pale yellow, many plants will produce a showy array of bright yellow flowers in later inflorescences.

Leguminosae

(Pea family)

Cassia fasciculata   Partridge pea (Plate 84)

During July and August in eastern and central parts of the state the showy yellow flowers of the partridge pea appear. Under some conditions of lush vegetative growth the flowers are almost hidden in the shadows of the many pinnately compound leaves. The name sensitive pea has also been applied to this plant, but its leaves are apparently not as sensitive to touch as are those of *Schrankia* which will be discussed later. The abundance of flowers on some plants is well shown in the accompanying picture as are the pea pod-like fruits in which small brown seeds are produced. This plant is found more often in disturbed soil areas than it is in prairie sod.

Leguminosae

(Pea family)

Desmanthus illinoensis   Prairie mimosa (Plate 85)

Prairie mimosa is one of the sensitive plants of prairie areas of most of the state where it grows in thickets, in ditches, and in rocky areas. Numerous heads of whitish flowers set on long stalks resemble pincushions. Flowers produce large numbers of flat, curved pods which often form brown balls when dry. The many bipinnately compound leaves with innumerable tiny leaflets fold together when the plants are disturbed. Unlike those of *Schrankia,* the stems of these plants are not barbed. Flowers may appear at almost any time of the growing season but probably are most abundant during July.

Leguminosae
(Pea family)

Desmodium illinoense   Tick trefoil (Plate 86)

One of several species of this genus found in Nebraska, tick trefoil grows to be four or five feet tall in good soil and is commonly
in flower during the last part of July and during August, sometimes into September.

Abundant purple flowers at the tips of several upper branches
are fairly conspicuous and are soon followed by the development
of a flattened, segmented, pea-like fruit which botanists call a loment. The sections of the loment often stick to the clothing as one
walks through weedy patches of this plant. This feature has led
to the name beggar-ticks for this plant. The three-foliolate leaves
have suggested the name trefoil as the common name for this
genus.

Leguminosae
(Pea family)

Glycyrrhiza lepidota   Wild licorice (Plate 87)

An almost strictly erect plant of the prairies, this plant grows to
a height of about three feet. In many seasons it grows well ahead
of the tall grasses among which it is found and thus is conspicuous.
Flowers of this plant have been described as white or yellow, but
usually they appear to be greenish yellow. While there are still
flowers in bloom some of the first flowers will have set fruit.
These are closely set with hooked prickles which soon turn brown.
It is often easier to recognize wild licorice in fruit than in flower as
the flower clusters often resemble those of *Astragalus canadensis*.
The glandular dotted leaflets of the wild licorice should serve to
distinguish it from an *Astragalus*. Also the flower clusters of the
licorice are from axillary shoots while those of *Astragalus canadensis* appear to be terminal on the plant.

There is only one species of this genus found in Nebraska. It is
from another species that the licorice of commerce is extracted.
The genus name literally means sweet root.

43

Leguminosae
(Pea family)
Lathyrus polymorphus   Showy vetchling (Plate 88)

When the novice comes upon the flowers of showy vetchling for the first time, the idea that a wild sweet pea has been found may flash through his mind. The common name indicates that this plant produces some of the most beautiful flowers to be found in this family. In spite of the meager number of flowers produced their beauty compensates for their scarcity. June is the month of their most frequent flowering although in some years they flower in May and July also. The purple standard, the light pink wings, and the darker pink keel are fairly typical of these flowers. The keel contains both the stamens and the pea pod-like pistil. The stamens may be found to have one stamen separated from the other nine near the end of the sheath of filaments, but all are united near the base of the filament sheath.

The stems are usually erect and about one foot tall. The leaflets arise in pairs and are often about an inch long but are fairly narrow and linear in shape. This species usually lacks tendrils at the ends of the leaves.

Leguminosae
(Pea family)
Lotus corniculatus   Bird's-foot trefoil (Plate 89)

Several bright yellow flowers in a ring borne in great profusion on decumbent plants from a few inches to two feet tall distinguish bird's-foot trefoil from other legumes. The individual flowers are pea-like but give a very striking bright yellow appearance to passengers flitting past these wayside flowers. These have been planted, apparently, along Interstate 80 in some places, and they seem to thrive in the median and shoulders even under mowing conditions.

The leaves are compound. There are three leaflets near the tip and two other stipule-like leaflets at the base. The common name is not derived from the leaves, however, but from the appearance of groups of dried pods.

The flowers are produced in greatest profusion during June and July.

Leguminosae
(Pea family)

Lupinus plattensis　　　　　　　　　　Nebraska lupine (Plate 90)

In the more arid sections of western Nebraska lupines grow in dry sandy soil. The leaves are composed of seven to nine leaflets attached at one place (digitately or palmately compound). Each leaflet is hairy on the underside and glabrous on the upper side. Hairs, which are whitish and appressed, clothe the petioles and stems of the plant as well as the flower stalk. This gives a silvery appearance to edges of the leaves and the stems.

The showy terminal raceme is covered with many blue flowers. The single flower cluster may be from one-fourth to one-half the height of the plant and appears as a blue spire among the surrounding vegetation.

This plant may be found in blossom during June and July and is fairly common in the western counties of the state.

Lupinus decumbens is fairly similar to the above species but has lighter blue flowers since they lack the dark streak on the standard which is the erect petal across the top of the flower. Bracts of the flowers are awl-shaped and may be retained for a short time. Also, the two parts of the calyx are of striking difference in length, whereas in the above species they are nearly equal.

Leguminosae
(Pea family)

Medicago sativa　　　　　　　　　　　　Alfalfa (Plate 91)

Although usually grown as a crop, alfalfa may occasionally escape from cultivation and appear in fence rows or ditches in the eastern part of the state. There are many variations in intensity of the flower color notwithstanding their usual description as purple. Some plants run to blue and others to pinkish hues. Often the first cutting of alfalfa is made before July, and usually it has been in flower before that time. Ripe pods are curved and more or less coiled together. Leaves are commonly trifoliolate (with three leaflets) and each leaflet has several sharply pointed teeth at the end. The plants are sparingly pubescent and are usually under three feet tall.

45

Leguminosae
(Pea family)
Melilotus officinalis   Yellow sweet-clover (Plate 92)

Yellow sweet-clover is one of two common sweet-clovers found in Nebraska. The bright yellow flowers, small trifoliolate leaves, and diffuse habit make these plants attractive. They are tall, much branched plants whose leaflets are elliptical or narrower, finely serrate except at the base, and about one inch long. The flowers are about one-fourth of an inch long, and the standard is not appreciably longer than the other petals. Blossoms are borne in spike-like racemes of considerable length. This species may come into blossom a week or ten days before white sweet-clover growing in the same habitat.

Melilotus alba, white sweet-clover, is very similar to yellow sweet-clover except that the plants may be taller, and the standard of the flower is usually longer than the other petals.

Flowering of both species occurs during most of the growing season but is most striking during June and July.

Leguminosae
(Pea family)
Petalostemon candidum   White prairie-clover (Plate 93)

Of all the prairie-clovers found in Nebraska the white and the purple are perhaps the most common in the eastern part of the state while the silky prairie-clover is the most common species of the sandhills.

White prairie-clover is a little more robust than the purple prairie-clover. The stems are stouter; the compound leaves have large leaflets or leaf segments (one inch or more) and are most numerous (five to nine). The cylindrical flower heads are larger and often come into flower sooner with the closely set flowers blossoming from the base of the flower head to the top. The species name, *candidum*, means white—thus the common name. These plants blossom in June and July.

46

Plate 68   Pennycress
Thlaspi arvense

Plate 69   Virginia stonecrop
Penthorum sedoides

Plate 70   Wild strawberry
Fragaria virginiana

Plate 72   Tall cinquefoil
Potentilla arguta

Plate 71   Silver-weed
Potentilla anserina

Plate 73   Sulfur cinquefoil
Potentilla recta

Plate 75   Lead plant
Amorpha canescens

Plate 74   Prairie rose
Rosa suffulta

Plate 76   Milk vetch
Astragalus canadensis

Plate 78   Ground plum
Astragalus crassicarpus

Plate 77   Bird's-egg pea
Astragalus ceramicus

Plate 79   Woolly locoweed
Astragalus mollissimus

Plate 80   Racemed milk vetch
Astragalus racemosus

Plate 81   Prairie milk vetch
Astragalus adsurgens

Plate 82   Large white wild indigo
Baptisia leucantha

Plate 83   Wild indigo
Baptisia leucophaea

Plate 84   Partridge pea
Cassia fasciculata

Plate 85   Prairie mimosa
Desmanthus illinoensis

Plate 86   Tick trefoil
Desmodium illinoense

Plate 87   Wild licorice
Glycyrrhiza lepidota

Plate 88   Showy vetchling
Lathyrus polymorphus

Plate 89   Bird's-foot trefoil
Lotus corniculatus

Leguminosae
(Pea family)
Petalostemon purpureum   Purple prairie-clover (Plate 94)

The slender, erect form of the purple prairie-clover may be observed in the prairies over most of Nebraska for almost a month before flowering begins. The terminal cylindrical heads of many minute flowers appear at the tips of erect branches. As Rickett (1966) suggests, because of the dominance of the red hues the flowers are not actually purple in color. Blossoming from the base toward the tip gives a very delightful appearance to the flowering head.

This species comes into bloom a week or so later than the white prairie-clover. These plants usually do not grow in thick stands but are common in prairies of rich soil. Flowering occurs during late June and July.

Leguminosae
(Pea family)
Petalostemon villosum   Silky prairie-clover (Plate 95)

From the eastern limits of the sandhills westward, this low spreading silky prairie-clover is at home. The numerous, often horizontal, flowering heads and an abundance of leaves give an aspect quite different from the preceding species of this genus. Its presence in almost clean sand instead of heavy sod is also surprising when one first observes this plant.

The leaves are odd-pinnately compound having many paired, oblong leaflets along the rachis and a terminal leaflet as well. The leaflets seem to grow with a silvery radiance reflected from the thick, villous pubescence.

Within a head, flowering progresses from the base to the tip; but the spread of flowers which are in blossom at one time covers more than half the length of the head. This gives an appearance of the whole head's being in flower at once in contrast to the blooming of the purple prairie-clover. The corolla is a light purple or pinkish purple and is often faded in the strong light from the sun and the sand. Flowering extends from July into August.

47

Leguminosae
(Pea family)

Psoralea argophylla   Silver-leaved scurf-pea (Plate 96)

The scurf-peas are also called psoralea (the initial *p* is silent), a convenient name since it is also the genus name. This psoralea can be found in slightly drier areas than the other species of *Psoralea* although it is often found along with them in good rich soil.

The silver-haired leaves distinguish this species whether it is in flower or not. The leaflets, commonly five in number, are digitately or palmately arranged.

The flowers, about one-third of an inch long and few in number, form groups on the ends of the terminal branches. Deep dark blue in color, they are the darkest blue of any of the local psoraleas or perhaps of any of our local flowers. Flowering may occur from June to August, following the flowering of both *P. esculenta* and *P. tenuiflora.*

Leguminosae
(Pea family)

Psoralea esculenta   Indian breadroot (Plate 97)

Indian breadroot grows up to two feet tall and is an erect plant, but its lateral branches extend as wide as the plant is tall.

The leaves are larger than those on the other psoraleas, and they typically have five digitate leaflets which are hairy, as are the stems. The compactly arranged flowers in the inflorescence are also covered with long hairs. Apparently many of the hairs arise from the sepals. The petals are pale, ranging from almost white to lavender, but are considerably larger than are those of other psoralea flowers. Blossoms are produced in June, but the rather large inflorescence is usually retained for another month.

These plants grow best in the prairie areas on well-drained soil. The root system descends from the lower end of an enlarged storage organ which develops two or more inches below the surface of the soil. One common name for this plant is Indian turnip because of the starch-filled, ball-shaped storage organ which was used as food by Indians and pioneers.

Leguminosae
(Pea family)
Psoralea tenuiflora   Many-flowered scurf-pea (Plate 98)

Perennial rhizomes of this scurf-pea produce erect, bushy, herbaceous plants from two to three feet tall in numerous pastures throughout the state. During June they are covered with racemes filled with small blue to violet flowers. The flower stalks extend above the leaves and this produces a colorful sight for the whole area. Each small pea-like flower produces a one-seeded, somewhat globular pod which does not normally open to release the seed.

The leaves are compound, having mostly three leaflets per leaf, but some of the lower leaves have five leaflets. The leaflets may be considerably over an inch long. They are strigose on the lower surface and glabrate on the upper surface.

In the past a common name of this plant was wild alfalfa, and fields of this scurf-pea have been mistaken for alfalfa. However, two quickly distinguishing features of these plants are the elongate racemes of the flowers and the green separate pods, which contrast with the heads of flowers and the brown, curved, flattened, clustered pods of the alfalfa.

Leguminosae
(Pea family)
Schrankia nuttallii   Sensitive brier (Plate 99)

Because of the curved spines along the stem, another common name used for this plant is catclaw sensitive brier. The leaves are pinnately compound with many pairs of small leaflets, which close when they are touched or when a person walks through a thicket disturbing the natural position of the stems—hence the name sensitive. These plants will grow in relatively poor soil, if it is well drained and dry. They are found mostly in the eastern and central part of the state and are often in flower during July or August.

The flowers are in small heads with stamens having purple or lavender filaments increasing the size of the ball-shaped cluster to the size of a marble.

49

Leguminosae
(Pea family)
Thermopsis rhombifolia   Prairie golden-pea (Plate 100)

In sandy soil of the western, particularly northwestern, part of the state these perennial plants are most spectacular during their flowering period in June and July. This species has been known by other common names such as prairie thermopsis, prairie bean, yellow pea, and false lupine. The genus name means having the appearance of the lupine. The species name is given for the rhomboid shape of the three leaflets which make up each leaf. At the base of each leaf there are two ovate stipules which are nearly the same size as the leaflets and make the leaves appear as though they had five leaflets. The stems are weaker than those of the lupine and they bear appressed hairs. As the plants dry out the foliage does not become black as it does in some legumes.

The brilliant yellow flowers are borne in terminal bracted racemes of several flowers. Each flower is short-stalked and has a calyx of sepals united at their bases but their free tips are green and long-pointed. The standard (erect petal) is rounded, yellow, and dark-grooved at the midrib. Other dark streaks radiate from the base of the standard. The oval wings (lateral petals) are also yellow and cover the upper side of the keel. From below, the keel is quite apparent and somewhat greenish along the midline on the outside. Within the keel, the stamens are separate from one another instead of being joined together by their filaments. The fruits are flat pods three to four inches long.

Leguminosae
(Pea family)
Trifolium pratense   Red clover (Plate 101)

Most of the red clover found in the state is grown in meadows as a crop, but it strays readily from domestication and may be found in a wide variety of habitats within the eastern part of the state. The trifoliolate leaves usually are mottled and comparatively large—at least larger than most other clover leaves. These are supported by long petioles and occur in clumps.

The red clover flowers are clustered together into a ball-shaped head. Each flower is fairly long and thus they are attended mostly during blossoming by bumble bees. The shades of color in the petals of different plants vary widely. Flowering occurs at almost any time during the growing season.

Leguminosae
(Pea family)

Trifolium repens   White clover (Plate 102)

Before the days of weed killers white clover was a very common plant in parks of eastern Nebraska and in the lawns of private homes and public buildings. It is not a particularly good plant for lawns because of its need for more water than is usually available during the summer months and its low tolerance for heat and short cutting.

The corollas, which are usually white, give their color to the flowers that are clustered into somewhat spherical heads. After the trifoliolate leaves develop for a time in the spring, the flowers begin to bloom and may continue until winter. Children often search areas of white clover to find four-leaf clovers, which are traditionally considered lucky.

Leguminosae
(Pea family)

Vicia villosa   Hairy vetch (Plate 103)

As with most of the vetches the terminal end of the leaf rachis continues as a tendril. This feature allows the vines to scramble over themselves and any surrounding vegetation. The vines display one-sided racemes of densely arranged flowers that present a mixed blue and white appearance. Hairy vetch is often grown for fodder, but it frequently escapes from fields to roadsides and fence rows where it may continue to flourish for a limited number of years. The greatest burst of flowering of escaped plants in eastern Nebraska occurs during June although sporadic blossoming may occur during any of the summer months.

51

Linaceae
(Flax family)
Linum sulcatum   Yellow flax (Plate 104)

The common name given here is not distinctive of this species since within Nebraska one may find two or perhaps three other species of yellow flax. Another name sometimes applied to this species is grooved flax, which is taken from the grooved condition of the stem between narrow wings. These plants are slender with very narrow leaves.

This species blooms during June and July with bright yellow flowers about an inch across. Usually a single plant will produce only a flower or two each day. Often the corolla is shed by afternoon.

The sepals are retained through the maturation of the fruit while in other species the sepals are shed when the fruit is first formed. The margins of all sepals are minutely toothed.

Oxalidaceae
(Wood-sorrel family)
Oxalis stricta   Yellow wood-sorrel (Plate 105)

Yellow wood-sorrel is found in most parts of the state and is in flower from April to October.

These plants are low growing, producing several flowers at one time. Most of the flower parts are yellowish including the five styles and the ten stamens. The flower stalks become reflexed in fruit but the capsules, which are hairy, remain erect.

The leaves, palmately three-foliate with obcordate leaflets, are subject to sleep movements at night. They have a sour watery juice but nevertheless are used for greens.

Oxalidaceae
(Wood-sorrel family)
Oxalis violacea   Violet wood-sorrel (Plate 106)

Violet wood-sorrel is most common in the southeastern part of the state but may also be found in the eastern half in secluded woody areas and in moist prairie areas.

52

It is known to produce flowers during the spring in May and June and on occasion to produce a second set of flowers without leaves during September or October. This has been observed on sandstone outcrops near Lincoln.

The flowers are often violet in color and exceed the leaves in height. The flower stalk arising from a scaly bulb may support either a single flower or a small group of flowers.

The leaves as in other wood-sorrels are palmately divided into three leaflets, each notched at its apex. They arise from the underground bulb instead of from a stem and may be somewhat more fleshy than those of the yellow wood-sorrel.

Geraniaceae
(Geranium family)
Geranium carolinianum   Carolina geranium (Plate 107)

The wild geraniums are often called crane's-bill because of the appearance of the ripe fruits after they have split open. The Carolina geranium is the only species that is at all common in the state. In eastern Nebraska the flowers are produced in April or May and the drying plants with fruits are seen in June. These rather short plants, usually found in well-drained, porous soil, have many leaves crowded together in a bushy branched habit that seldom exceeds a foot in height. The leaves are palmately dissected with nearly linear ultimate lobes. Flowers are borne in a terminal branched cluster at about the same height as the leaves.

Zygophyllaceae
(Caltrop family)
Tribulus terrestris   Puncture vine (Plate 108)

Puncture vine is an introduced weed which is found throughout the state. In dry areas and waste places, it may become a serious hazard because of the hard, spiny fruits which are capable of puncturing bare feet or bicycle tires. It is listed as a primary noxious weed in Nebraska. The plants are prostrate, bearing opposite pinnately compound leaves. By branching freely at the nodes they form a rather complete ground cover. The flowers, which are short

53

stalked and pale yellow, are borne in the axils of the leaves. The fruits resulting from a single flower produce five burs, each with two spines. Flowers and fruits are found on the plant during the hot summer months of June, July, and August.

Polygalaceae
(Milkwort family)

Polygala alba   White milkwort (Plate 109)

Among the early grasses on partially tree-covered slopes of northwestern Nebraska may be found these slender, slightly branched herbs arising about a foot tall from woody rootstocks.

The flowers are borne in narrow spikes arranged spirally to irregularly on the stalk. A great treat is in store for anyone who, with a good hand lens, starts investigating these tiny flowers. The two prominent lateral white flower structures are two of five sepals, the other three being small and often greenish. The three petals are connected to each other and to the stamen tubes. The stamens have anthers with but a single pollen pore at the end instead of the two slits that open along the side of the anthers in most other flowers. Flowering occurs during June and July.

Euphorbiaceae
(Spurge family)

Euphorbia corollata   Flowering spurge (Plate 110)

This family of latex-producing plants is very large with many diverse kinds of plants, from cactus-like succulents to woody trees and small prostrate herbs. They have greatly reduced flowers, several occurring within tiny, often glandular, involucres. The glands are sometimes accompanied by white appendages that give the involucre the appearance of a flower.

The flowering spurge has the largest appendages of any species of *Euphorbia* in our region. It grows in pastures, waste places, and roadsides in the eastern part of Nebraska and is a showy plant when in blossom. About three feet tall, with whorls of several redivided branches making up the inflorescence, this plant may be very conspicuous for its many "flowers."

54

Plate 90   Nebraska lupine
Lupinus plattensis

Plate 91   Alfalfa
Medicago sativa

Plate 92   Yellow sweet-clover
Melilotus officinalis

Plate 93   White prairie-clover
Petalostemon candidum

Plate 94   Purple prairie-clover
Petalostemon purpureum

Plate 95   Silky prairie-clover
Petalostemon villosum

Plate 96   Silver-leaved scurf-pea
Psoralea argophylla

Plate 97   Indian breadroot
Psoralea esculenta

Plate 98   Many-flowered scurf-pea
Psoralea tenuiflora

Plate 99   Sensitive brier
Schrankia nuttallii

Plate 100   Prairie golden-pea
Thermopsis rhombifolia

Plate 101   Red clover
Trifolium pratense

Plate 102   White clover
Trifolium repens

Plate 103   Hairy vetch
Vicia villosa

Plate 104   Yellow flax
Linum sulcatum

Plate 105   Yellow wood-sorrel
Oxalis stricta

Plate 106   Violet wood-sorrel
Oxalis violacea

Plate 107   Carolina geranium
Geranium carolinianum

Plate 108   Puncture vine
Tribulus terrestris

Plate 109   White milkwort
Polygala alba

Plate 110   Flowering spurge
Euphorbia corollata

Plate 111   Snow-on-the-mountain
Euphorbia marginata

This is a perennial plant with the annual stems arising from deep roots. The stems bear many alternately arranged, nearly sessile, hairless, oblong leaves. The lowest leaves are often dried or already shed by flowering time which in Nebraska is August.

Euphorbiaceae
(Spurge family)

Euphorbia marginata   Snow-on-the-mountain (Plate 111)

Another species closely related to the flowering spurge is snow-on-the-mountain. There are many differences between this species and the previous one. In this species the showy parts of the plant are the white-edged bracts. Although the involucres have glands and white appendages accompanying them, they are not nearly as conspicuous as in the previous species. The inflorescence is not as highly branched as in flowering spurge, but the large bracts make this species look even more dense.

The plants grow each year from a tap root into a somewhat flat-topped herb from one to three feet tall. The alternate leaves may be three inches long. They are conspicuously in bloom during July and August, especially in the eastern half of the state. Some home owners have used these plants in their flower gardens.

There are many other species of Euphorbia growing in Nebraska, some of which are closely related to poinsettia and several ground spurges, but none are particularly attractive. *Euphorbia esula,* the leafy spurge, has linear leaves and heart-shaped bracts which are yellowish and somewhat conspicuous.

Balsaminaceae
(Touch-me-not family)

Impatiens capensis   Spotted touch-me-not (Plate 112)

Most of the species of *Impatiens* are called touch-me-not, jewelweed, snapweed, or balsam. There are two species of the genus found in Nebraska of which this one is the more common and more highly colored. The scientific name which authorities formerly used for this plant is *Impatiens biflora.*

This is an annual which grows in moist shady areas protected from strong winds. The stems are rather weak, watery, and somewhat translucent. Juice from crushed stems is a good antidote for poison ivy. The stem and leaf surfaces are smooth and dew often accumulates as drops which glisten in morning light, thus the name jewelweed.

The leaves are oval with irregular but not deeply notched edges. From the axils of the leaves irregular flowers hang suspended on slender stalks which are attached at about the middle of the tapering tubular corolla that ends as a curved or reflexed spur. Flowers are colored yellow to orange, with red to reddish brown spots.

The common name, touch-me-not, refers to the sudden opening of the fruits which upon being touched eject seeds with some force. Flowering and fruiting occur during June and July.

Malvaceae
(Mallow family)

Callirhoe involucrata   Purple poppy-mallow (Plate 113)

The flowers of this plant are not abundant and, except in a few areas, the plants do not grow in extensive patches. If these two conditions were reversed, many people of the state would nominate this plant as our state flower. The color of the flower is very intense and the colloquial name wine-cup has been used in some localities.

The plants are decumbent and low-growing, bearing leaves that are palmately divided into several narrow and redivided lobes. Flower stalks bear but one flower, which is closely subtended by three small, narrow bractlets. The calyx is five-parted and hairy as is also the flower stalk. The calyx lobes are similar in texture to the bractlets but are longer. The crimson or purple petals tend to overlap laterally in the formation of a beautiful cup-like corolla. The central structure of the flower is composed of the whole set of stamens with more or less united filaments surrounding the many styles. Flowering occurs in June.

Another species commonly found blooming in the state at the same time is the pale poppy-mallow, *Callirhoe alcaeoides*. It has pink to white flowers, no involucre, and a more upright habit of growth.

Malvaceae
(Mallow family)
Hibiscus militaris   Marsh mallow (Plate 114)

The pithy clustered stems up to five feet tall, which are relatively smooth and soft, bear variously shaped leaves. The lateral divergent basal lobes of the leaves may or may not be present and in some stands are almost completely absent. The plants grow in shallow water along streams and occasionally will cover the entire area of a roadside borrow pit if there is a nearly constant supply of water. This species occurs in the eastern part of the state.

Flowers are produced on short stalks from the axils of upper leaves during August. White or pinkish corollas open in the morning and display the brilliant red at the center of the flower for only a brief period of time during midmorning. Again, as is typical of the family, the many stamens form a column with the anthers divergent along the sides for most of its length. The stamen column surrounds the united styles except for the five cream-colored stigmas which protrude separately on style branches at the outer end of the column. The flower bud opens by the spreading of the five green sepals which remain united at their bases.

Malvaceae
(Mallow family)
Hibiscus trionum   Flower-of-an-hour (Plate 115)

As a small, hairy weed with bright, upturned flowers, this plant is delightful to observe from a distance and to examine closely with a hand lens. These flowers are noted for being open only a few hours of the day. The petals are white, cream, or occasionally yellowish with bases making up a deep purple center against which the yellow anthers and divergent bright red, capitate stigmas show off as a miniature spectacle when magnified.

Found in flower in waste places as early as mid-June, by August these plants have developed in stubble fields left from the harvest of wheat or other small grains. Until September they may be common around feed lots. The leaves are deeply divided into three lobes which are redivided into blunt or obtusely pointed segments.

57

Malvaceae
(Mallow family)

Malva rotundifolia   Running mallow (Plate 116)

As a weed this plant has spread to many parts of the United States as well as to scattered areas of this state. It grows close to the ground in an ascending habit and produces leaves on long petioles with roundish blades which are only faintly lobed. Flowering is from June to September.

Features which distinguish this plant from *Malva neglecta* are the hairiness at the base of the petals and the comparative length of petal and sepal. The light-colored hairs do not show up well against the lighter-colored base of the petals. The sepals are not shown in the photograph, but they are only slightly shorter than the petals while in *M. neglecta* the petals may be twice as long as the sepals. Several species of *Malva* have a broad, shallow notch at the end of each lilac-colored petal.

Malvaceae
(Mallow family)

Cheyenne 6/85

Sphaeralcea coccinea   Scarlet mallow (Plate 117)

On sandy soil the scarlet mallow grows as a low herbaceous plant producing a profusion of brick red flowers during June and July. The color of the petals varies considerably in nature as well as in photographs. Shown in the picture are flowers that appear salmon-colored. The center of the flower has a column of bright yellow stamens which flare out into a whorl of anthers at the end just beneath the stigmas.

The leaves are alternately arranged on the stem and are deeply parted with several narrow segments. The stems vary from light green to a yellowish color. Of interest to those looking closely at the stems and leaves will be the stiff stellate hairs which cover much of the surface of the plant and collect many grains of sand.

Hypericaceae
(St.-John's-wort family)
Hypericum perforatum   Common St.-John's-wort (Plate 118)

These plants grow erect and in certain localities become quite closely entwined to the extent that on walking through a dense stand one's progress is severely impeded. The plants are usually less than three feet tall and have linear-oblong, opposite leaves which bear dots along their margins and are sessile.

The flowers are produced in profusion at the top of the plant, making a bright yellow display. The five sepals lack dots and are half as long as the petals. The five petals are each notched along one side from beyond the middle to the apex and are black-dotted on the margin. The stamens are aggregated, usually into three groups (sometimes five), with many stamens in each group. Most flowering occurs in June and July in various parts of the state, but sporadic flowering may occur until September.

Violaceae
(Violet family)
Viola pedatifida   Prairie violet (Plate 119)

As in many violets the crown of the plant is developed on an erect rhizome. The leaves and flower stalks arise from the crown, which is near soil level. Leaves are divided into several lobes and some or all of the lobes are further divided into two or more ultimate segments which are nearly linear. Conspicuous petal-bearing flowers are borne on stalks which are usually taller than the leaves. The peduncle then curves downward and supports the hanging flowers from the top. The two top, two lateral, and one lower petal are similar to those of other violets except in technical details. It is the lower petal that is continued into a spur or sac. It usually has hairs around the edge of the sac similar to the hairs near the base of the lateral petals shown in the photograph. The two lowest of the five stamens are provided with nectar-bearing appendages which extend into the sac. Nonopening (cleistogamous) flowers are also produced on ascending peduncles.

59

Many other species of violets occur in Nebraska, some with white flowers and others with yellow. Some have true stems above ground from which both leaves and flowers arise. Anyone interested in these should consult a technical manual.

Loasaceae
(Loasa family)

Mentzelia decapetala   Sand lily (Plate 120)

The common name of this plant is rather misleading with regard to *lily* but is very appropriate with regard to *sand,* for to my knowledge it is found only on sandy soil. It generally grows in the western part of the state but does occur eastward in suitable habitats to Niobrara and even farther east. It is to be found in blossom from June to September. It is considered to be a short-lived perennial or perhaps a biennial in some places. The plant grows erect with alternate leaves which have hooked hairs, coarse teeth, and a wavy margin.

The flowers are differentiated at the tip of each lateral branch in the upper part of the plant. Petals are nearly white and the center is cream-colored to yellowish. The flowers open in the evening and remain open all night but are closed during most of the day.

Cactaceae
(Cactus family)

Coryphantha vivipara   Purple cactus (Plate 121)

This cactus has also been called ball cactus. It may be found in several of the northwestern counties of Nebraska, usually blooming during May and June. The ball-shaped body has many cylindrical tubercules. These are terminated by a set of outward pointing, dark-colored spines, often five or more in number. Surrounding these is a whorl of many grayish spines. Bright purple flowers make this cactus particularly attractive. Each flower is subtended by many brownish sepals which are fringed with white. The petals, which are the bright purple parts, are slender and very pointed. Many yellowish anthers fill the center of the flower but do not equal the white styles in height.

60

Cactaceae
(Cactus family)
Opuntia fragilis   Brittle cactus (Plate 122)

Since the terminal segments readily break away, this plant is known as brittle cactus. It grows along on the ground forming mats of plump to cylindrical or slightly flattened spiny segments. The spines vary in number from one to seven at a place and each segment is covered by many spine-bearing areas. This is in contrast to ball cactus in which there are spines only at the tip of the segment. The flowers are yellow with darker centers. The anthers are numerous and, when shedding pollen, are lighter in color. The fruits are dry, nonedible, and covered with spines.

Cactaceae
(Cactus family)
Opuntia tortispina   Prickly-pear (Plate 123)

Broad, flat, fleshy stem segments characterize this cactus which is popularly known as prickly-pear. The plants root freely at the joints and may form mats of various sizes, some of which may be several feet in diameter. One to four spines and many minute smaller bristles which form a cushion about them are produced at each spine-bearing area. These bristles often cause inflammation in tissues in which they are embedded. The spines are reddish at the base and at the tip and may be up to an inch in length. Flowers occurring during June and July are sulphur yellow and about three inches in diameter. The fruit has no spines and is fleshy and edible. It also gives this plant its common name.

Lythraceae
(Loosestrife family)
Ammannia coccinea   Tooth-cup (Plate 124)

If you are searching for something among marshy or lowland areas, you may on occasion come upon this small plant among other vegetation such as fog-fruit *(Phyla lanceolata)*. As found in Nebraska, tooth-cup is often scarcely a foot tall. The leaves are narrow, opposite, sessile, and auricled, almost clasping the stem.

61

The flowers are not conspicuous, being sessile in the axils of the leaves or borne two or three in an axil. The calyx is tubular or cup-like and four-toothed, the whole cup being about one-sixth of an inch long. A small appendage at each sinus of the cup accounts for the common name. The four petals are purplish and are shed rather soon after the flower opens, which is during May and June.

Lythraceae
(Loosestrife family)
Lythrum alatum   Winged loosestrife (Plate 125)

When you next come upon the winged loosestrife as you wander among swamp or wet ditch vegetation, stop for a second and look at the flowers with your hand lens. Along the slender wand-like winged branches, the flowers are usually produced singly in the axils of the leaves. If you find as you examine the flowers closely that some have not six petals, but five or seven, you need not be too surprised. Look again at the flowers. Are there always the same number of stamens as there are petals? One plant may have long styles with terminal stigmas far beyond the flower or the calyx tube. Another plant with flowers bearing long exserted stamens may have short styles allowing the stigma to get barely to the edge of the tubular calyx. Suggestions have been made that dimorphic flowers, such as these, increase the chances for cross-pollination.

Onagraceae
(Evening-primrose family)
Epilobium angustifolium   Fireweed, great willow-herb (Plate 126)

This is a tall plant with purple flowers which are produced in terminal clusters. The buds at the tip of the flower cluster droop down against the stem. They rise to a horizontal position at the time of flowering and become somewhat ascending as the elongated fruits mature. The flower parts occur in fours: four sepals, four petals, usually eight stamens, and a four-parted stigma. The sepals, petals, and anthers are purple. The stamens come out of the flower first and are followed by the style with its four-lobed stigma opening after the pollen has been shed. It has been suggested

Plate 112   Spotted touch-me-not
Impatiens capensis

Plate 113   Purple poppy-mallow
Callirhoe involucrata

Plate 114   Marsh mallow
Hibiscus militaris

Plate 115   Flower-of-an-hour
Hibiscus trionum

Plate 116   Running mallow
Malva rotundifolia

Cheyenne 6/85

Plate 117   Scarlet mallow
Sphaeralcea coccinea

Plate 118   Common St.-John's-wort
Hypericum perforatum

Plate 119   Prairie violet
Viola pedatifida

Plate 121   Purple cactus
Coryphantha vivipara

Plate 120   Sand lily
Mentzelia decapetala

Plate 122   Brittle cactus
Opuntia fragilis

Plate 123   Prickly-pear
Opuntia tortispina

Plate 124   Tooth-cup
Ammannia coccinea

Plate 125   Winged loosestrife
Lythrum alatum

Plate 126   Fireweed
Epilobium angustifolium

Plate 127   Scarlet gaura
Gaura coccinea

Plate 128   Common evening-primrose
Oenothera biennis

Plate 129   Evening-primrose
Oenothera rhombipetala

Plate 130   Tooth-leaved
evening-primrose
Oenothera serrulata

Plate 132   Poison hemlock
Conium maculatum

Plate 131   Water hemlock
Cicuta maculata

that this condition favors cross-pollination of flowers. The seeds each bear a tuft of hairs which allows them to be carried considerable distances by air currents. The leaves are alternate, nearly entire, and up to eight inches long. Flowering occurs during any of the summer months.

Onagraceae
(Evening-primrose family)
Gaura coccinea   Scarlet gaura (Plate 127)

Along roadsides and in dry prairies, scarlet gaura may be found throughout Nebraska. It is fairly common and blossoms in greatest abundance in June and July. Each plant produces several spreading and ascending branches with stems up to a foot tall. The leaves are simple and narrowly tapering with minute teeth along the margins.

The plants have terminal spikes of many buds from which arise small flowers. The blossom is long and slender with four scarlet petals and the four sepals reflexed back against the floral tube. There are usually eight stamens and the stigma is four-lobed. The fruits in this genus are not as long as in other genera of this family.

Onagraceae
(Evening-primrose family)
Oenothera biennis   Common evening-primrose (Plate 128)

Oenothera is the genus from which this family derives its common name. There are many species of this genus in Nebraska, whose flowers vary from white through lavender to bright yellow. Petals of these flowers are large, making the blossoms showy and conspicuous.

The most common of the tall yellow-flowered evening-primroses in this state is *O. biennis*. The four yellow petals are rounded to lobed and are overlapping, forming a cup-shaped flower. These are produced in a terminal spike having many bracts. Though probably most common in the central part of the state, they are fairly well distributed and may be found in blossom from as early as June through the rest of the summer and until frost. Older plants may become shrubby.

The leaves are long and narrowed to a point and are sessile on the stem; their margins are slightly wavy and have small teeth.

Onagraceae
(Evening-primrose family)
Oenothera rhombipetala   Evening-primrose (Plate 129)

Of all the common roadside species, the most beautiful one is *O. rhombipetala*. The specific name is derived from the rhomboid shape of the pointed yellow petals. The erect habit of this plant combined with prolific flowering of the terminal inflorescence provides a spectacular specimen from almost every plant. Look for these along roads in the sandhills and western part of the state during June and July.

The calyx turns down with the opening of the flower. The central parts of the flower show the style divided into four long, narrow stigmas and usually eight stamens whose anthers are also long and narrow. All of these parts are yellowish.

Onagraceae
(Evening-primrose family)
Oenothera serrulata Tooth-leaved evening-primrose (Plate 130)

Oenothera serrulata is found in well-drained or dry soil over most of the state. The habit of the plant is low and often branched. The leaves are narrow, long, keeled, and toothed along the margin.

The flowers are yellow and are set in the axils of the upper leaves. Before the flowers open, the calyx is strongly four-ribbed with a pink color between the green ribs. Opening of the flowers reveals the deep yellow stamens, usually eight in number. The stigma is not divided but the end of the style is enlarged and rounded into four lobes. The ovary is inferior as in all members of this family and often appears as the thickened stalk of the flower below the corolla. This species may be found in flower from May to July.

Umbelliferae
(Parsley family)
Cicuta maculata   Water hemlock (Plate 131)

In the family Umbelliferae there are many plants used in cooking but this and the next species are among the most poisonous plants of our local flora.

64

The tuberous roots clustered at the base of the stem are the most poisonous part of this plant. Even a small part of one of these roots is likely to be fatal to man and other mammals. Death comes after violent convulsions and finally complete paralysis any-where from fifteen minutes to eight hours after ingestion of the root material.

It grows in wet places and around the springs at Victoria State Park. The leaves of this plant are compound but the leaf segments are not as finely divided as in poison hemlock *(Conium)*.

The flowers are very similar to *Conium*, but the fruits are not winged nor are they particularly poisonous.

The stems of this plant are splotched with purple lines similar to those of *Conium*. Early in the season the roots may be pulled up and eaten by cattle feeding on the above-ground parts. Thus most loss of livestock from water hemlock occurs in the spring of the year.

Umbelliferae
(Parsley family)

Conium maculatum   Poison hemlock (Plate 132)

Conium is from the Greek word meaning hemlock. Poison de-rived from this or similar plants in Roman times and earlier was referred to as hemlock. The species name *maculatum* is given to plants which are splotched or spotted with color. In this case the tall, erect, much-branched stems which are a light green color have irregular reddish or purple areas. These plants appear to be bienni-als; they are glabrous herbs with three or four times pinnately compound leaves and many small white flowers in relatively flat-topped inflorescences usually called umbels.

The large compound leaves have leaflets which are ovate in out-line and the ultimate segments are pinnatifid. The petioles are di-lated and sheathing at the base.

The many minute flowers have white petals but no sepals and are borne in umbels which are one to three inches broad. Flower-ing is prominent in June and July. Fruits are distinctly winged.

Least poisonous is the root and most poisonous are the fruits and seeds. Death does not always result from eating the seeds since stimulants can be used to advantage. The jaws are not locked shut as with *Cicuta* nor are convulsions a usual symptom. Coma, slowing of the heart, and finally paralysis of the organs of breathing bring about a relatively quiet death if the patient is not treated.

Umbelliferae
(Parsley family)
Daucus carota   Wild carrot, Queen Anne's lace
(Plate 133)

This plant has been introduced from Europe. It becomes a weed in roadsides and dry waste places in many parts of the state. It is usually a biennial with a deep fleshy root. Stems grow to heights of one to three feet and are mostly erect. The lower leaves are two or three times pinnate with the ultimate segments often pinnatifid. The upper leaves are smaller with fewer divisions, but the segments are nearly linear. The inflorescence is on a long slender stalk. The bracts of the involucre are divided into linear lobes. The whole umbel of flowers is usually from two to four inches across. Individual flowers occur on very slender short pedicels in smaller clusters, many of which make up the whole inflorescence. The flowers are usually white although pinkish-flowered forms are known. The small central flowers of the compound umbels are often dark red or purple instead of white. The ribs on the small mature fruits are winged and bristly. Flowering is most abundant in July but may continue into September.

Umbelliferae
(Parsley family)
Heracleum lanatum   Cow parsnip (Plate 134)

Cow parsnip is a large-leafed plant which grows in northern latitudes and is found most frequently in Nebraska in wet places along the Niobrara River drainage area. This perennial plant has leafy stems and the leaves are ternately compound, with broad leaflets which are palmately lobed, pubescent, and sharply serrate. The petioles are very wide-spreading above the base. The flowers are white or sometimes pinkish. Fruits are broadly oval, up to one-half inch long, finely pubescent, and somewhat notched at the summit. Flowering in Nebraska occurs during June and July.

Umbelliferae
(Parsley family)

Lomatium foeniculaceum   Whiskbroom parsley,
hairy parsley (Plate 135)

One of the early-flowering Umbelliferae, this plant may be seen in flower during April. It may be found growing in rocky, well-drained prairies or woodland areas of the state.

The plants are perennial, arising from somewhat tuberous, upright roots without any stem to speak of. The petiole is strongly sheathing. The leaves, which are ternate, have each part further dissected three times into linear ultimate segments often less than one-fourth inch long. The yellow flowers occur in umbels which lack involucral bracts. The bractlets of the small divisions of the umbel are conspicuous, united to the middle or beyond, and occur on one side of the umbellet. The fruits are oval and glabrous with broadly winged lateral ribs.

Another similar species, *Lomatium orientale*, is of similar growth habit but has white flowers instead of yellow.

Umbelliferae
(Parsley family)

Lomatium nuttallii   Dog parsley (Plate 136)

One of the spring flowering members of the parsley family, the dog parsley, is found in Nebraska only in our most western counties, Sioux and Scotts Bluff. This seems to be the eastern extent of its range, which extends through Wyoming and Utah. Higher altitudes and drier habitats seem to enable this plant to thrive.

A visit in May to the Scotts Bluff National Monument should reveal these bright yellow flowers growing from low, closely clumped stems. The flower clusters do not greatly exceed the foliage of the plant in height. The leaves are pinnate or bipinnate with pointed linear leaflets. These arise from a stout stem-like structure which is densely covered by old leaf sheaths. The fruits are strongly flattened and oblong. Lateral ribs of the fruits are broadly winged while other ribs may be sharp or more or less winged.

67

Umbelliferae
(Parsley family)
Pastinaca sativa   Parsnip (Plate 137)

The cultivated parsnip belongs to the same species as the escaped form shown in the picture but is designated as a different variety. The fusiform root, as is usual with all parsnips, develops the first year and seems to thrive after being frozen during the winter. In the second year of growth the whitish root produces a plant five feet tall with large pinnate leaves and several scattered umbels of greenish yellow flowers. The stems are large, hollow, and grooved on the outer surface. The broad leaflets are variously lobed or merely dentate. The compound umbels whose many flowers have five yellow petals and no sepals appear greenish because of the relatively large green pistil in the center of each flower. The mature fruits are flat, glabrous, and strongly winged.

These plants have been cultivated since ancient times, and as escapes they are to be found along roadsides in the eastern third of Nebraska.

Umbelliferae
(Parsley family)
Sium suave   Water parsley, Water parsnip (Plate 138)

As a common plant growing in water or wet places in the northern half of the United States, this plant resembles the water hemlock. It has been accused of being poisonous in three accounts according to Kingsbury (1964), but in his judgment these are not convincing cases.

These are coarse, erect perennials occasionally attaining a height of three feet in our region. The white-flowered umbels are similar to those of *Cicuta,* but the plant differs in having strongly ribbed stems and once-pinnate leaves with linear or narrow lanceolate leaflets having serrate margins and acuminate tips. As with many plants, however, the leaves that develop under water are highly dissected. The umbels have many rays up to an inch or two long. The fruits are flattened and glabrous but have nearly equal, corky ribs. The plants usually flower during July, but blooms occasionally appear as late as October.

Ericaceae
(Heath family)
Monotropa uniflora   Indian pipe (Plate 139)

These startling small, white plants, rare within the state but sometimes found in rich, moist woods, are saprophytic or perhaps parasitic. As fleshy plants lacking chlorophyll, they are usually less than ten inches tall. The stems bear white scales or bracts instead of leaves. A single nodding, odorless flower develops at the apex of the stem. There are four or five separate white petals and eight to ten stamens in the flower. As the fruit matures it changes to an erect position on the stem as may be seen in the photograph. With drying, the parts of the plant become black. This may also come about as a result of frost in late-developing forms. The photograph was taken on the east side of the Missouri River on October 1.

Primulaceae
(Primrose family)
Dodecatheon pauciflorum   Shooting star (Plate 140)

Along streams in western Nebraska, these small, beautifully flowered plants may be found in their prime in May or June before the native grasses in low meadows begin their growth. In good seasons, acres of meadow land may be colored by the purple petals of these exquisite plants. The leaves are basal, forming a rosette, and from these the scapes bearing a terminal umbel of several flowers arise. The calyx is cleft and reflexed. The corolla has a very short tube with the petals reflexed and spreading upward as the flower faces downward. There is often a heavy purple wavy line at the edge of the cup of the corolla with lighter coloration above the line.

Primulaceae
(Primrose family)
Lysimachia ciliata   Fringed loosestrife (Plate 141)

In dissected upland meadows or hayfields one may often find the fringed loosestrife in the border area between a mowed section

69

and a wooded ravine. Also, in grassy or weedy draws, one may find this upwardly branching plant in blossom in early to mid-July. The light yellow petals of the flower on plants up to three feet tall and the persistent sharp sepals around the developing ovary are characteristics which often identify this plant on first sight. Confirmation of the identity may be assured by the axillary placement of the solitary flower stalks and by the ciliate nature of the petioles as can be seen in the accompanying photograph. The leaf margins are also often seen to be minutely ciliate.

Primulaceae
(Primrose family)
Lysimachia nummularia   Moneywort (Plate 142)

This creeping, mat-forming plant may be used in shaded streetside parkings of cities as a ground cover which will tolerate moderate mowing. The opposite, rounded leaves have short petioles. The bright yellow flowers with parts mostly in fives often are not numerous, but they are sufficiently large to make the ground cover attractive in the last half of June.

Primulaceae
(Primrose family)
Lysimachia thyrsiflora   Tufted loosestrife (Plate 143)

One of our marsh plants frequent in swamps and along the edges of ponds, particularly in the sandhills, is the widespread tufted loosestrife which is found across the continent from California to West Virginia and northward as far as Quebec and Alaska.

This perennial has simple, smooth to slightly villous stems arising from the water in a swampy area, as pictured along the Platte River in central Nebraska. The sessile leaves are lanceolate and may be as much as five inches long.

From the axils of the middle leaves arise the stalked racemes of yellow flowers crowded into club-shaped clusters. Individual flowers usually have six parts in each whorl. The yellow corolla is deeply divided into six narrow petals, less than one-fourth inch long, which often have small black dots. The stamen filaments are yel-

Plate 133   Wild carrot
Daucus carota

Plate 134   Cow parsnip
Heracleum lanatum

Plate 135   Whiskbroom parsley
Lomatium foeniculaceum

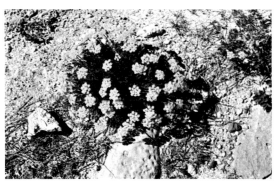

Plate 136   Dog parsley
Lomatium nuttallii

Plate 137   Parsnip
Pastinaca sativa

Plate 138   Water parsley
Sium suave

Plate 139   Indian pipe
Monotropa uniflora

Plate 140   Shooting star
Dodecatheon pauciflorum

Plate 141   Fringed loosestrife
Lysimachia ciliata

Plate 142   Moneywort
Lysimachia nummularia

Plate 143   Tufted loosestrife
Lysimachia thyrsiflora

Plate 144   Prairie-gentian
Eustoma grandiflorum

Plate 145   Closed gentian
Gentiana andrewsii

Plate 146   Downy gentian
Gentiana puberulenta

Plate 147   Buckbean
Menyanthes trifoliata

Plate 148   Prairie dogbane
Apocynum sibiricum

Plate 149   Common periwinkle
Vinca minor

Plate 150   Sand milkweed
Asclepias arenaria

Plate 151   Swamp milkweed
Asclepias incarnata

Plate 152   Showy milkweed
Asclepias speciosa

Plate 153   Sullivant's milkweed
Asclepias sullivantii

low and about twice as long as the petals. This gives the whole flower a fuzzy yellow appearance. Flowering occurs from May to July.

Gentianaceae
(Gentian family)

Eustoma grandiflorum   Prairie-gentian (Plate 144)

The deep purple petals of the prairie gentian form large bell-shaped flowers not too unlike a tulip in general shape. The prominent yellow anthers are supported on filaments which become fused with the petals in the lower part of the flower. Even more conspicuous than the anthers is the large two-lobed yellow stigma. Usually less than two feet tall, the plants have opposite, elliptical leaves and a glaucous covering over much of the surface of the vegetative structures. Flowers are in best display in July and early August. These plants seem to be avoided by cattle and thus may be found to be fairly prominent in some fields used exclusively for pasturing. The plant seems to be a biennial or at least an overwintering annual. In a fair-sized pasture some plants will be found to have white flowers and occasionally some have pinkish flowers.

Gentianaceae
(Gentian family)

Gentiana andrewsii   Closed or bottle gentian (Plate 145)

The name closed gentian is derived from the flowers whose corolla remains nearly closed at the apex even when it is in full bloom. The name bottle gentian arises from the shape of the corolla which is generally that of a flask or bottle. As with many gentians, the corolla is blue or purple, sometimes apparently changing with age. At the tip of the stem are several flowers borne in clusters. The stems are usually erect and bear opposite sessile leaves. Moist edges of hayfields are the commonest locations for these plants, and September is the month of most abundant flowering.

71

Gentianaceae
(Gentian family)
Gentiana puberulenta   Downy gentian (Plate 146)

The downy gentian is a slightly smaller plant (about one foot tall) than the previous species (about two feet tall). It is similar in that it blossoms prominently in September but, unlike the closed gentian which looks as if it will open in blossom momentarily but never does, the downy gentian is seldom seen until the blossoms are open. There is similarly a dense cluster of flowers near the top of the plant. Each flower lacks a stalk but bears five blue to lavender petals whose lobes are considerably longer than the corolla tube which has two-cleft plicate folds between each lobe. The stamens do not join to the corolla as they do in many other species but remain free. The perennial stems are slender and stiff, and the sessile leaves are opposite and successively at right angles to each other (decussate).

Gentianaceae
(Gentian family)
Menyanthes trifoliata   Buckbean (Plate 147)

The buckbean is one of Nebraska's rare plants. It is shown in the picture as it was photographed near Brewster, Nebraska, in early May. Found in bogs and swamps in northern North America, the plants are herbaceous but perennial by a creeping rootstock. Leaves are mostly basal but technically alternate, with long sheathing petioles and three leaflets (hence the specific name). The oval leaflets are two to four inches long and are thick, glabrous, and entire. In the flower the sepals are short, but the petals are one half or more inches long and bearded on the inner side. Most petals are white but may be pinkish; they are united at their base into a short tube.

Apocynaceae
(Dogbane family)
Apocynum sibiricum   Prairie dogbane (Plate 148)

The inclusion of such a weedy plant as this in a wild flower book may evoke consternation in the minds of many people who

72

have weeds and wild flowers in separate, well-defined categories in their minds. However, to some fortunate people who can look upon beauty of form and symmetry without the prejudice of knowing the menacing attributes of certain plants, this might be a plant worthy of inclusion.

The flowers of this plant are small, but the white petals, green sepals, and almost stylized axillary branching immediately characterize the genus *Apocynum*. People who do not recognize it by name are curious to know about the plants that have milky juice, yet by their flower are different from the milkweeds of the genus *Asclepias*.

Another almost indistinguishable species is *Apocynum cannabinum*, the Indian hemp. The lack of, or length of, the petiole of the opposite leaves of the main stem and the green bracts and white flowers distinguish *A. sibiricum* from *A. cannabinum*, with its petioled main stem leaves, scarious bracts, and greenish white flowers. Both species may have variations in leaf shape and in the presence or absence of a mucronate tip on the leaf.

The fruits are narrow follicles up to four inches long in *A. sibiricum* and up to eight inches long in *A. cannabinum*. The tuft of hairs on each seed is much longer in the second species.

Apocynaceae
(Dogbane family)

Vinca minor   Common periwinkle (Plate 149)

Vine-like ground covers, which are evergreen, easily controlled, and yet suitable for growth in Nebraska are hard to find, yet *Vinca minor* fits these requirements very admirably. In certain areas when left alone it will escape and grow as a wild vine. It is useful in landscaping under dense shade trees, on steep banks, or in other unmowed areas. It is not a strong competitor and will not produce a lush growth when used in mowed or dry areas.

The firm, dark green, glossy, oblong leaves remain until covered with snow and then are again present in the spring as soon as the snow melts.

The flowers are blue with united petals forming a short narrow tube at the base, in which the stamens are located. Blossoms occur during April and May and then sporadically the rest of the year.

73

Asclepiadaceae
(Milkweed family)
Asclepias arenaria   Sand milkweed (Plate 150)

The sand milkweed is found mostly in the sandy parts of Nebraska including the central and western parts of the state. It may be found up to the edges of blowouts such as the one pictured west of Cairo, Nebraska.

The plants are ascending but usually not erect even though the stems are rather stout. The leaves are somewhat squarish in outline, often having broadly rounded distal ends which may be slightly notched.

The flowers are greenish because of the reflexed petals, but the hoods are white and longer than the anthers. From the hood emerges a white, slender horn which turns inward and ends above the stigma.

The most common time of year for flowering is during July although it may occur from June to September.

Asclepiadaceae
(Milkweed family)
Asclepias incarnata   Swamp milkweed (Plate 151)

Plants with milky juice and opposite leaves often characterize the milkweeds. This milkweed is found in swamps and sloughs and may grow to four or five feet in height, mingled with other vegetation.

The lanceolate leaves are opposite and each pair is set on the stem about 90° from the pair above and below. They do not exude as much milky juice when broken from the stem as some milkweeds growing in dryer places. The leaves are numerous and pointed at the end, and their veins angle upwards as they branch from the midrib. There are hairs on the leaf but casual observation may not detect them.

The pink or rose-colored flowers occur in several umbels with many flowers per umbel. The petals are most deeply colored with the hoods lighter, and the horns and stigma are lighter yet. July and August are the months in which most flowering occurs.

74

Asclepiadaceae
(Milkweed family)

Asclepias speciosa   Showy milkweed (Plate 152)

The showy milkweed may be found blossoming during June and July in wet prairies, valleys, and plains in central and western Nebraska. It is called the showy milkweed because of the length and divergence of the hoods. These may be a half inch or more in length and are light purple or pink. Each hood has a horn which points inward from below the middle of the hood. The reflexed petals spread outward showing the delicate purple lines which give the flowers their purplish appearance.

The differences in leaves between this species and the common milkweed are not great, but in the showy milkweed the leaf bases are likely to be somewhat heart-shaped. Petioles in both species are very short. The dense soft pubescence on the leaves gives them a velvety feel.

Asclepiadaceae
(Milkweed family)

Asclepias sullivantii   Sullivant's milkweed (Plate 153)

Anyone knowing the common milkweed would think this plant was the same until the smooth hairless stems and leaves come to his attention.

The purplish hoods with similarly colored horns are well shown in the picture while the petals which reflex back against the flower stalk can hardly be seen. The light-colored center of the flower over which the horns point is the stigma.

The milkweed flowers are usually so plentiful during the summer months that we often neglect to stop and investigate their intricacies. The five-sided structure in the center of the flower has been referred to as the stigma. It is more than that since it includes the anthers as well. There are five anthers, each with two pollen sacs. The neighboring pollen sacs from adjacent anthers are connected together by a structure which has been called a translator. Ordinarily in walking around on the flowers, insects will pull a leg up between adjacent anthers, catching the translator and pulling the pollen sacs out. The two bags of pollen, called pollinia, contain

75

sticky pollen grains. Small flies are frequently unable to pull out either the pollinia or their leg and they die, caught in one of nature's traps.

Asclepiadaceae
(Milkweed family)

Asclepias syriaca   Common milkweed (Plate 154)

The common milkweed is often collected in the fall of the year by those who desire natural winter bouquets of the seed pods, or follicles, as they are technically called. I presume that almost every child has blown seeds from the boat-like follicle and watched them float away in the breeze, dangling on that surprising hairy type of parachute and defying gravity over long journeys in the air currents. One of the marvels of nature to me as a youth was how that inner surface of the follicle could be so soft and smooth while the outer was relatively rough with protruding points. How many adults, though, have ever stopped to investigate the central part of a single milkweed flower to see the origin of the two potential pods. Many, I suppose, have never stopped to consider the connection between the fragrant flowers of the summer and the pods of autumn.

The under surfaces of leaves have short soft hairs. Leaf shape may be oval or somewhat pointed, and the petioles are very short. The large, many-flowered umbels may be attracting insects from June through August. The flowers are purplish with noticeably hairy flower stalks of the same color.

Asclepiadaceae
(Milkweed family)

Asclepias tuberosa   Butterfly weed (Plate 155)

The butterfly weed is very successful in attracting butterflies during its flowering period from June to September. Once a plant is seen in full flower no further explanation is necessary for its common name.

Other features may need some explanation, however. The leaves on this plant are alternate on the stem instead of opposite. When

76

one is pulled from the plant, no milky juice runs from the leaf. A discerning observer may see that a clear drop of fluid is collecting. This has been shown to be a clear type of plant latex instead of milky latex.

The orange-colored flowers are different from the purple or greenish flowers of other species, but the flower parts are unmistakably those of a milkweed. Some plants show an almost red flower.

This perennial plant is attractive enough to make a nice specimen in home plantings. Propagation should be by seed rather than transplanting. Once established, it needs scarcely any care, but it is sensitive to herbicides. These plants seem to be disappearing from the prairies.

Asclepiadaceae
(Milkweed family)

Asclepias verticillata   Whorled milkweed (Plate 156)

To those who live in the prairie provinces of this continent the whorled milkweed is one of the most commonly known milkweeds. In the western plains area it is considerably less frequent if not absent entirely.

The leaves, which are very narrow, are usually set in whorls on the slender stem. On the lower part of the stem the leaves may be scattered. The umbels of white flowers are lateral and terminal. Each flower is relatively small but very typical of a milkweed. In the dry prairies the whorled milkweed grows among the early-season grasses. These plants are usually not seen before they flower because of their slender stems and narrow leaves.

Asclepiadaceae
(Milkweed family)

Asclepias stenophylla   Narrow-leaved milkweed (Plate 157)

This species and the following one belong to a part of the genus *Asclepias* in which the horns of the flowers are not conspicuous and often the hoods are considerably modified.

The narrow-leaved milkweed has a slender to medium-sized stem which is usually erect and up to three feet tall. The upper

77

leaves tend to be alternate but the lower leaves are opposite.

The hoods are whitish and usually appear to be three-lobed at the top. Some authors call one of the lobes a horn.

The pods on this milkweed are erect.

Asclepiadaceae
(Milkweed family)
Asclepias lanuginosa   Woolly milkweed (Plate 158)

These plants grow among the other vegetation on the prairie. Stems may become two feet tall or more; they bear lanceolate leaves which are hirsute on both surfaces. The species name means downy and the common name is derived from the same feature.

Umbels grow singly at the summit of each stem where they remain erect. In each small flower the purplish hoods are not as long as the anthers, and they are appressed to the central structure of the flower. This species is in the section of the genus whose flowers lack horns. The flowers appear purplish and they blossom during May and June.

Asclepiadaceae
(Milkweed family)
Asclepias viridis   Spider milkweed (Plate 159)

One of the rarest milkweeds of the state belongs to a part of the genus *Asclepias* which is characterized by having the petals simply spread when in blossom rather than reflexed. The flowers of this species are probably among the largest of the genus. Each flower in the rather diffuse umbel is more than an inch in diameter. The plant is ascending and it bears pairs of oblong leaves. Within the flower the most conspicuous parts are the purple and white hoods. The central part of the flower consisting of fused anthers and stigma is typical of other milkweeds. It may be found in flower during June and July. The picture was taken in the southwestern part of Richardson County on June 27.

Plate 154   Common milkweed
Asclepias syriaca

Plate 155   Butterfly weed
Asclepias tuberosa

Plate 156   Whorled milkweed
Asclepias verticillata

Plate 157   Narrow-leaved milkweed
Asclepias stenophylla

Plate 158   Woolly milkweed
Asclepias lanuginosa

Plate 159   Spider milkweed
Asclepias viridis

Plate 160　Field bindweed
Convolvulus arvensis

Plate 161　Bush morning-glory
Ipomoea leptophylla

Plate 162　White-flowered gilia
Gilia longiflora

Plate 163　Standing cypress
Gilia rubra

Plate 164　Creeping phlox
Phlox andicola

Plate 165　Wild blue phlox
Phlox divaricata

Plate 166   Prairie phlox
Phlox pilosa

Plate 167   Virginia waterleaf
Hydrophyllum virginianum

Plate 168   Miner's candle
Cryptantha celosioides

Plate 169   Blueweed
Echium vulgare

Plate 170   Hairy puccoon
Lithospermum caroliniense

Plate 171   Narrow-leaved puccoon
Lithospermum incisum

Plate 172   False gromwell
Onosmodium molle var. *occidentale*

Plate 173   Fog-fruit
Phyla lanceolata

Plate 174   Woolly verbena
Verbena stricta

Plate 175   Ground ivy
Glecoma hederacea

Plate 176   Motherwort
Leonurus cardiaca

Convolvulaceae
(Morning-glory family)
Convolvulus arvensis   Field bindweed (Plate 160)

This species is one of the most troublesome weeds of our state. It scrambles over many other plants and uses them for support by twining around them so tightly that their stems are deformed. The vines are often very extensive and are hard to kill because of the very deep root system.

The alternate leaves are hastate, giving a roughly triangular outline to the whole leaf. Before flowering occurs the leaves of this plant may be distinguished from those of wild buckwheat by the convex margin between the widest part and the leaf tip while the margins are concave in the wild buckwheat vine.

The large funnelform corollas give a clue to its family affiliation. Usually they are white but often are somewhat pink, and they are about an inch across. The stamens are barely exserted from the corolla tube and the style is elongate, bearing two linear, divergent stigmas. Flowering may occur from May through September.

Convolvulaceae
(Morning-glory family)
Ipomoea leptophylla   Bush morning-glory (Plate 161)

A ragged shrub-like plant with narrow gray green leaves, the bush morning-glory grows to over three feet in height and spreads twice that far laterally.

The rose purple corolla is funnel-shaped and spreading at the outer end. It arises out of a short green-striped calyx and extends two to three inches. These conspicuous flowers are borne in abundance on certain branches.

The most surprising aspect of this plant is its extensive underground development. Below the surface at varying distances the root expands into a tremendous fleshy root up to one and a half feet in diameter and three feet long. It may have the shape of an overgrown turnip, being broadest near the top and tapering down to normal root size again at the bottom.

79

The bush morning-glory grows best in dry sandy soil of our sandhills but also extends into western Nebraska. It may blossom from May through July.

Polemoniaceae
(Phlox family)
Gilia longiflora   White-flowered gilia (Plate 162)

One of our common flowers of the sandhills of western Nebraska is the white-flowered gilia. It may bloom any time in the growing season from May to September. The photograph was taken about the middle of July at the Valentine National Wildlife Refuge.

This species is glabrous with the erect stems and panicle branches displaying a shining green color. The leaves are pinnatifid, with long, narrow segments. The plants often are about two feet tall and have elongated tubular flowers about two inches long. The glabrous sepals are united into a calyx tube with short teeth at the apex of the tube. The tubular corolla is white and has the stamens inserted unequally near the top of the tube. As indicated in the illustration the corolla may drop, leaving the elongated slender style extending from the short calyx tube.

Polemoniaceae
(Phlox family)
Gilia rubra   Standing cypress (Plate 163)

Although the cultivated form of this plant is not too often grown in flower gardens of this state, it is known to have escaped cultivation in Lancaster County as indicated in the photograph.

The plants are biennial and may grow to three feet in height although the plants illustrated are closer to two feet tall. The leaves are closely set and finely dissected into filiform segments on the unbranched stems.

The prominent tubular corolla is a vivid red. Stamens are barely exserted from the corolla tube, but the style and its white stigma protrude farther. The cluster of flowers at the upper end of the stem provides a spectacular sight during the summer months.

Polemoniaceae
(Phlox family)

Phlox andicola   Creeping phlox (Plate 164)

Formerly botanists called this plant *Phlox douglasii* but rules of priority seem to indicate that it should be called *P. andicola.* Thus to continue calling it Douglas' phlox seems inappropriate, yet creeping phlox applies to a number of cultivars from several different species.

The leaves, which are up to one-half inch long on this decumbent plant, are awl-shaped, rigid, and usually longer than the internodes of the stem. They may be devoid of hairs or again may be slightly white-woolly. The plants attain a height of two to four inches and generally occur in tufts.

The flowers have tubular corollas which are longer than the calyx; they usually exceed one-half inch in length and are generally white although one may occasionally find pink forms. Flowers are mostly solitary but may occur in clusters of up to five flowers. Ordinarily a covering of white flowers over one of these small tufted plants comes from solitary flowers on closely growing stems.

For the most part, these plants appear in dry soil of the plains in western Nebraska and bloom from May to July.

Polemoniaceae
(Phlox family)

Phlox divaricata   Wild blue phlox (Plate 165)

There are so many common names attributed to this early flowering herb of our woodlands that it is easier to call this plant by its scientific name. The common name sweet William is really misapplied when used with this genus but it is used so often some people now call this plant sweet William phlox.

These small plants which grow up to about a foot tall are variable as to the intensity of blue in the petals. Some plants exhibit very pale blue flowers. The early flowering of the wild blue phlox is a popular attractive feature. The photograph was taken southeast of Weeping Water, Nebraska, in late April. In some years in favored habitats it flowers in early April.

81

Unlike the upper leaves seen in the illustration, the basal leaves of this plant are more rounded and without acute tips, sometimes being described as ovate with obtuse tips. There are also to be found at the base of these plants leafy prostrate shoots which act as runners.

Polemoniaceae
(Phlox family)

Phlox pilosa   Prairie phlox (Plate 166)

Unlike the previous species, the plants of this species are found in prairies rather than in woodlands, and they bloom in summer rather than in spring. They have no leafy prostrate shoots. Their flowers are likely to be red purple or some hue of pink or even white rather than blue. The corolla tubes are usually pubescent rather than glabrous. Their leaves are lance-shaped and pointed and do not have the more rounded leaves at the base of the stems.

Similar to the previous species, these plants are erect, herbaceous, and up to a foot or more tall. Their flower clusters are similar in bearing several to many flowers which are loosely arranged or wide-spreading. The calyx is about half as long as the corolla, and the teeth of the calyx are about as long as the calyx tube. These plants are found in the eastern part of the state.

Hydrophyllaceae
(Waterleaf family)

Hydrophyllum virginianum   Virginia waterleaf (Plate 167)

The Virginia waterleaf is one of our rare plants, but it thrives rather well in flower gardens in eastern Nebraska. In cultivation it produces an abundance of foliage in comparison to the number of flowers.

The leaves are broadly triangular in outline and are pinnately divided almost to the midvein into five or more segments. Some segments may be lobed, and each is terminated by an acute tooth.

It bears its flowers with the first flush of leaf growth during May or June, and its fruits remain conspicuous until midsummer. Flowers have united petals which are lavender and somewhat bell-shaped. The filaments, which are longer than the corolla by a considerable amount, are the conspicuous structures bearing nearly black anthers. The calyx and inflorescence generally are hairy.

Boraginaceae
(Borage family)
Cryptantha celosioides   Miner's candle (Plate 168)

Growing in the dry gravel of Smiley Canyon of Sioux County, this plant was photographed during the last week of June. It seems to thrive in the plateau country in the northwestern part of the state. It grows from a stout root and is either a biennial or a short-lived perennial. The stems are characteristically single yet branched in the upper part of the plant to produce a narrow paniculate inflorescence in which the flowers tend to be crowded into glomerate clusters. The individual flowers are tubular with the stamens included in the tube and the crests at the end of the tube forming a white or yellow corona.

The plant is covered with sharp hairs, particularly on the stems. The lower leaves are narrow or spatulate and alternate. Such strictly erect plants as these seem picturesque in such a dry area.

Boraginaceae
(Borage family)
Echium vulgare   Blueweed (Plate 169)

Blueweed, also known as blue devil and viper's bugloss, has been described as a bad weed of open fields but has not been listed in the bulletin, *Nebraska Weeds.* There are a few places in Lancaster County where it grows in abundance, but it seems not to be a serious invader. The plants are armed with fiercely stiff and sharp spines making them hard to handle.

The deep blue flowers are bilaterally symmetrical (irregular) rather than radially symmetrical (regular) as are most of the other flowers of this family. The funnelform corolla with its open throat is composed of five segments with rounded and spreading lobes. The stamens arise from the base of the flower, which may be purplish instead of dark blue, and are exserted well beyond the co-

83

rolla and somewhat unequal in length. The style is filiform and is cleft into two stigmas at the tip. The calyx is segmented into usually five hairy, narrow, pointed parts which are considerably shorter than the blue corolla.

The leaves are sessile and narrow or lanceolate and coarsely ciliate.

Boraginaceae
(Borage family)
Lithospermum caroliniense   Hairy puccoon (Plate 170)

The hairy puccoon and the hoary puccoon, *Lithospermum canescens,* are somewhat similar plants but differ in our area in several respects. As the common names indicate this plant has rough, hairy (even hispid) leaves while the hoary puccoon has soft, appressed hairs which may give a whitish cast to the foliage. Hairy puccoon grows in a form which appears to be somewhat less crowded than the clustered, compact heads of the hoary puccoon. The flowers of the hairy puccoon are about one-third larger than those of the hoary puccoon.

These are short plants little more than one foot tall when in blossom. They have bright yellow or orange yellow corollas which are tubular with, commonly, five rounded segments and crests at the top of the tube. The stamens are borne within the tube, the anthers being nearly sessile.

Depending on local conditions flowering may occur as early as April and continue into June.

Boraginaceae
(Borage family)
Lithospermum incisum   Narrow-leaved puccoon (Plate 171)

The fimbriate margins of the thin yellow petals are the most characteristic structures of the narrow-leaved puccoon. The plants are similar in size to the other puccoons and also have appressed hairs similar to those of the hoary puccoon. The narrow leaves, however, help to distinguish this plant from the other puccoons when not in flower.

84

The genus name of these plants comes from the Greek, meaning stone seed, because of the hard fruits, often called nutlets, which these plants produce. In this species the nutlets vary from a shiny white to buff.

This species, similar to *L. caroliniense,* is distributed widely over the entire state and may be found in blossom from April to July on dry prairie soil.

Boraginaceae
(Borage family)
Onosmodium molle var. *occidentale*   False gromwell
(Plate 172)

It is unusual to find a flowering plant that becomes sexually mature before the flowers open, yet such is the case for this plant, the false gromwell. The style is exserted and the anthers are dehiscent before the corolla matures. The corolla is tubular with pointed lobes and the calyx is divided into unequal narrow segments. The flowers appear white.

The nutlets are smooth and shining without a collar. Mature plants, up to two feet tall, have leaves which are coarse with heavy vascular bundles. These slant away from the midrib and curve toward the tip of the leaf.

The plants are common in prairies and along streams in the central part of the state.

Flowering occurs from May through July.

Verbenaceae
(Vervain family)
Phyla lanceolata   Fog-fruit (Plate 173)

These plants, formerly known as *Lippia lanceolata,* are to be found growing in moist soil of meadows, ditches, and streams in eastern Nebraska and less commonly westward. This separate genus, *Phyla,* is sometimes used to segregate some of the herbaceous species from the tropical and subtropical American woody species in the genus *Lippia.*

85

The stems are green, nearly glabrous, slender and weak, becoming procumbent and rooting at the nodes. Erect stems are simple or only slightly branched, bearing lanceolate leaves that have more than four teeth on each edge. Toothing extends below the middle of the leaf.

The flowers are borne in heads or short spikes supported by elongating peduncles which arise from the axils of the leaves. The corolla is unequally four-lobed. The calyx is short and two-cleft and the fruit consists of two one-seeded nutlets. Flowering progresses from the lower to the upper flowers in the head. The months of flowering are June, July, and August.

Verbenaceae
(Vervain family)
Verbena stricta Vervain or woolly verbena (Plate 174)

The stem of *Verbena stricta* is strictly erect, hence the specific name. It is usually simple but it may be branched above, particularly within the inflorescence. The whole plant is soft to the touch because of the presence of soft, whitish hairs. The stem is somewhat four-angled, reflecting the insertion of pairs of opposite leaves on alternating sides of the stem at each succeeding node.

The oval leaves have serrate margins, short petioles, and prominent vascular bundles. Ordinarily they are erect and spreading.

The spikes bear many sessile, densely arranged, bracted flowers that vary from purple to lavender and almost blue. Flowering proceeds from the base of the spike upwards. The flowers are irregular in that the two upper petals of the united corolla are narrower and more deeply separated than are the three lower petals. The flowers, although small, are about twice the size of those of the blue verbena, *Verbena hastata,* and four times the size of those of the white verbena, *Verbena urticifolia.*

Labiatae
(Mint family)
Glecoma hederacea Ground ivy (Plate 175)

This plant, introduced from Europe, is one of the most pestiferous weeds of the flower garden and lawn. It will grow vegeta-

86

tively beneath snow and spread in all directions during the winter. It has other interesting names such as creeping Charlie, gill-over-the-ground, and runaway robin.

The leaves of the ground ivy have long petioles and rounded leaf blades with round-toothed margins (crenate). Prostrate stems root at the nodes.

Flowers occur from March through May. They are typical of the mints, with a blue or violet corolla and growing in small axillary clusters which surround the stem. The corolla is two or three times longer than the five unequal calyx segments, and the end of the corolla tube, which is called the limb, is two-lipped. The upper lip is erect and two-lobed or barely notched; the lower lip is three-lobed with the middle lobe broad and notched while the side lobes are small. There are four stamens, the upper pair being much longer than the lower pair; they are not exserted from beneath the upper lip of the corolla. With the aid of a good hand lens one may see that the two anther sacs are divergent from the tip of the filament.

Labiatae
(Mint family)

Leonurus cardiaca   Motherwort (Plate 176)

This perennial plant grows as an occasional weed in waste places. Its stem is stout, strict, slightly pubescent, and branching with a few straight ascending axillary shoots. The plants grow up to five feet tall bearing thin leaves which are partially cleft—the lower ones with two to five segments or lobes and the upper leaves with three lobes. The lobes are pointed at the tip and dentate along the sides. Numerous flower clusters occur at each of the upper nodes with many flowers in each cluster which entirely surrounds the stem. The calyx teeth are somewhat spreading and are nearly as long as the tube of the pink or white corolla. The anther sacs are parallel (not divergent) in these flowers. Flowers appear from June to September.

Labiatae
(Mint family)
Monarda fistulosa   Wild bergamot, horse-mint (Plate 177)

One of the tallest mints with large flowers is *Monarda fistulosa.* It is commonly called monarda as well as the other two common names listed. Its height of four feet compares with the giant hyssop, *Agastache scrophulariaefolia,* which is about five feet tall but whose flowers are small and in elongate spikes. The flowers of this plant are borne in somewhat rounded heads about an inch in diameter exclusive of the lavender floral tubes which may protrude about an inch. The floral tube is divided as in most mints into two limbs. The upper limb is tipped with a cluster of fairly conspicuous hairs.

The leaves are petiolate and somewhat hairy. They are lanceolate in shape with blades tapering into the petiole at the base. The edges of the leaf blades are unevenly dentate. These perennial plants grow throughout the state and flower in late June and sporadically through the summer.

Labiatae
(Mint family)
Monarda pectinata   Lemon monarda (Plate 178)

There are at least three species of *Monarda* in Nebraska which have this general aspect: the flowers are in compact clusters entirely surrounding the stem at its upper nodes. In this species the flowers are pink or white and are not spotted with purple. The calyx teeth are narrowed to short stiff bristles at their apex. The bracts at the nodes are nearly glabrous.

These annual plants are found on dry plains areas of the state and often are of quite low growth habit—less than one foot tall. In more favorable locations such as at Enders Dam their height may be up to two feet and the stem may be branched many times at the base, producing a sizable plant. They are found in flower during any of the summer months.

Monarda citriodora, lemon-mint, has pink or white flowers spotted with purple in the throat. The calyx teeth are similar to the above species. The bracts at the nodes are slightly hairy. This species is also an annual.

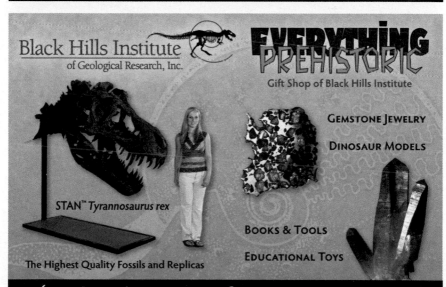

2009 Summer Events Schedule

Brought to you by the City of Hermosa
Email: twnhrmsa@custercountysd.com

Jun 21   Southwest Dakota 4-H Rodeo

Aug 13-16   Custer County Fair

Sep 26-27   Custer State Park
           Buffalo Roundup Arts Festival

Sep 26-27   Custer State Park
           Buffalo Roundup

COME WORSHIP WITH US!

Our Saviors Lutheran Church
255-4662

St. Michael's Catholic Church
343-3541

United Church of Christ
255-4503

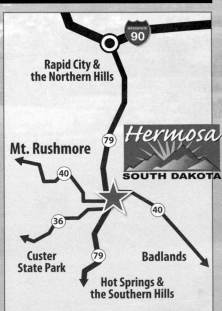

Rapid City &
the Northern Hills

INTERSTATE 90

Mt. Rushmore

79

Hermosa
SOUTH DAKOTA

40

40

36

Custer
State Park

79

Badlands

Hot Springs &
the Southern Hills

Let's Rodeo! *It's South Dakota's Official State Sport*

The Spanish term "rodeo" was first used in approximately 1834 by wranglers who gathered after a cattle roundup to celebrate with informal horsemanship and roping competitions. Legendary champions Russ Madison, Casey Tibbs, Paul Tierney, and Marvin Garrett became famous cowboys who exemplify the Dakota Territory rodeo tradition.

From Little Britches and 4-H, to High School and the PRCA— modern rodeo action comes in all shapes and sizes. Calf roping, bull riding, steer wrestling, barrel racing and bronc riding have become staples of this truly American sport. Cowboys and cowgirls from all over the nation compete in local professional rodeos, several of which have been honored as the best events of their kind in the nation.

For a genuine Old West South Dakota experience, take in a Rodeo!

NLBRA Spearfish – April 4-5

NLBRA Rapid City – April 18-19

GPIRA Crazy Horse Rodeo/Crazy Horse Memorial – June 19-21

4-H Southwest SD Rodeo/Hermosa – June 21

SD High School Rodeo Finals/ Belle Fourche – June 24-28

PRCA Black Hills Roundup/ Belle Fourche – July 2-4

SDRA Wall – June 9-11

PRCA Days of '76/Deadwood – July 21-25

PRCA Central States Fair/Rapid City – August 22-30

SDRA Oelrichs – September 12-13

PRCA Black Hills Stock Show/ Rapid City – January/February, 2010

Guide to Abbreviations:

GPIRA – Great Plains Indian Rodeo Association

PRCA – Professional Rodeo Cowboys Association

SDRA – South Dakota Rodeo Association

NLBRA – National Little Britches Rodeo Association

DAKOTA GOLD®

& Prairie Rose® COLLECTION

by Wheeler®

Rings, Earrings, Lockets, Pins, & Pendants...

For Gifts as Unique as the Beautiful Black Hills, visit one of the dealers listed below!

Note: Jewelry shown may may not be actual size

Monarda punctata, spotted horse-mint, has pale yellow flowers with purple spots in the throat. The calyx teeth are triangular or lanceolate. This species is perennial.

Labiatae
(Mint family)

Nepeta cataria   Catnip (Plate 179)

Around farmsteads of this state catnip seems to be one of the most common herbs. Often escaping into nearby woodlots or to pasture areas under the shelter of sumac, this is one of the most persistent plants, lasting years after any sign of habitation has disappeared.

The soft, whitish hairs give a distinct gray color to the whole plant, which may grow to a height of three feet by the time of flowering. The vegetative part of the plant is readily recognized by its odor which seems particularly attractive to cats—thus the common name. The erect plants bear opposite leaves which are distinctly toothed.

The flowers are numerous in several terminal spikes but are not conspicuous because of their small size and their white or light bluish color. Flowering occurs from late summer to fall.

Labiatae
(Mint family)

Prunella vulgaris   Heal-all (Plate 180)

These perennial plants are unbranched and small, usually less than one foot in height; they produce flowers in a terminal head or short spike.

If one examines a flower carefully with a hand lens the four stamens will be found to be sheltered by the upper lip of the corolla. Each filament seems to be divided near the tip into two teeth with the lower branch bearing the anther. The anthers occur more or less in pairs. The pollen sacs are not parallel and may be found to be discharging pollen at different times. The lower lip of the corolla is divided into three parts. The large central one is toothed around the margin and the smaller lateral segments have

89

entire margins. The bracts are green or purple with ciliate margins and short-toothed, sharp tips.

The leaves of a plant are few and petiolate and have margins which are entire or barely toothed. The stems of this plant are covered with many soft hairs. Found more commonly in the eastern part of the state, the plant is an introduction from Europe and now grows in waste places throughout most of North America. It may be found in flower during almost any part of the growing season from May to October.

Labiatae
(Mint family)

Salvia azurea   Blue sage, blue salvia (Plate 181)

As one of the typical perennial forbs in prairies and pastures, this mint may appear early in the season in the vegetative condition and yet not flower until the last half of July with occasional flowering continuing into September.

During its early vegetative growth it may be recognized by its keeled, recurved leaves which are petiolate and only slightly toothed. The upper leaves may be nearly linear in contrast to the lanceolate leaves produced on the lower portions of the plant.

The large, blue, two-lipped corolla is very conspicuous once flowering starts. The floral tubes are elongate and the lower lip is much larger than the horizontal or ascending upper lip of the corolla. The flowers are clustered into crowded whorls producing a spike-like inflorescence. Occasionally one may happen upon a variant bearing white flowers. Shown here is one of the bluest-flowered forms I have seen. Ordinarily the blossoms are a lighter blue.

Labiatae
(Mint family)

Teucrium canadense   Wood sage (Plate 182)

Growing as perennials in moist and usually somewhat protected habitats these plants occur in clumps as shown in the illustration. The plants have an erect habit of growth and often develop single, unbranched stems.

90

The leaves of this species are lanceolate to ovate and distinctly serrate. The leaves have short petioles and are covered with hairs. The lower surface of the leaf is covered with fine, grayish white pubescence, while the upper surface has dense, appressed pubescence.

The flowers are in a cylindrical, terminal, spike-like inflorescence six to ten inches long. The calyx and bracts have short, appressed pubescence. The corolla is over one-half inch long. Its four upper lobes which appear lavender with purplish throats are nearly equal in length and there seems to be no upper lip. The lower lobe is much larger. The stamens are exserted between the upper, smaller lobes, and the pollen sacs of the anther lie together. Flowering may occur from June to September with July furnishing the greatest flush of blossoms.

Solanaceae
(Potato family)
Datura stramonium   Jimson weed (Plate 183)

Jimson weed is a rank annual herb of waste places and is also commonly found around barnyards in eastern Nebraska. It is the natural source of the narcotic stramonium. The widely branching, glabrous stems bear large, alternate leaves which are variously and coarsely toothed and lobed. Flowers are regular and solitary in the axils of the leaves or branch stems. The calyx is tubular and angular with five small sharp-pointed lobes. Large white corollas each bear five acuminate lobes on the edge of the funnelform tube. Five stamens remain included within the corolla tube and their filaments are adnate to near the middle of the tube. The ovary, which bears a two-lobed stigma, later develops into a spiny, erect capsule about two inches long. When mature the capsule splits open by four valves releasing numerous flattened seeds.

Solanaceae
(Potato family)
Physalis heterophylla   Ground-cherry (Plate 184)

Sometimes the name clammy ground-cherry is applied to this species. Many specialists have admitted this to be a very confusing

91

genus with many varieties and forms attributed to this and other species of the genus. This ground-cherry is not noticeably clammy nor particularly pubescent. Most varieties of this species have dentate leaves but these leaves seem to have entire margins that are only slightly undulate. For those interested, U. T. Waterfall worked on this genus in 1958; he recognized more forms and varieties, not fewer.

Despite the scientific arguments it is comforting to the novice that he can recognize a ground-cherry almost anywhere he finds it. The stems may be branched relatively little and erect or may be much branched and decumbent. The leaves and the pubescence may be various, but the medium-sized, solitary flowers which arise from the leaf axils on distinct peduncles show many characteristics of this family. When one learns to recognize the older flowers with the enlarging calyx and knows of the berry within the mature inflated calyx, he hardly ever forgets the ground-cherry.

Solanaceae
(Potato family)

Solanum rostratum   Buffalo-bur (Plate 185)

Most lovers of wild flowers have taken the trouble to look at the flowers of the potato and the tomato, and it is no surprise to them that the buffalo-bur belongs to the same family, although some might not have suspected this prickly plant of belonging to the same genus as the potato.

The yellow flowers and yellowish spines on the stems and leaves denote this as a pestiferous weed which is found in almost every barnyard and many waste places and on disturbed ground throughout the state. These plants bloom from May to September, with many clusters of from three to fifteen flowers. After blossoming, the spiny calyx encloses the berry. The anthers are yellow with one often larger than the others. A hand lens will reveal the stellate pubescence throughout the plant.

The horse-nettle, *Solanum carolinense,* is also quite a common weed, but the five united petals are white to pale blue instead of yellow. The berry is yellow when ripe and is not enclosed in a spiny calyx. The five yellow anthers are quite conspicuous when the flowers are mature.

The black nightshade, *Solanum nigrum,* is found frequently throughout the state. Its leaves are nearly entire, its flowers are small with white petals, and its fruits are small and black at maturity.

Scrophulariaceae
(Figwort family)

Bacopa rotundifolia   Water-hyssop (Plate 186)

The round-leaved water hyssop may be found occasionally on muddy banks or muddy lake margins in the state. Flowering occurs from June to September. The flowers are white with yellowish throats. The corolla is almost equally five-lobed. The four stamens appear to be attached to the corolla at its throat, and the anthers before dehiscence are blue. The style is divided at the tip into two capitate stigmas which are well separated from each other, each one appearing to subtend an anther.

The young stems are pubescent and may extend for some distance. The rounded leaves have several distinct veins arising from the base.

One may find these plants in the drainage basin areas of Fillmore County. The photograph of this species was taken northeast of Fairmont, Nebraska, during late July.

Scrophulariaceae
(Figwort family)

Agalinis tenuifolia   Slender gerardia (Plate 187)

These small, angled-stemmed herbs are characterized by opposite, linear, entire leaves having only one main vein. They are usually somewhat roughened on the upper surface. The flowers are axillary to the upper leaves. The pedicels are filiform, divergent, and less than one inch long. The calyx is regular, forming a tube with very short lobes. The purple corolla is irregular; the tube bulges somewhat beyond the calyx and has five lobes, the upper two of which arch over the four stamens and are nearly erect. The lower pair of stamens is longer. The flowers are produced in late summer and each lasts but one day.

93

Scrophulariaceae
(Figwort family)

Linaria vulgaris   Butter-and-eggs (Plate 188)

This plant, originally introduced from Europe, is locally common in eastern Nebraska along roadsides, fence rows, and waste places. As a perennial herb, spreading by short rhizomes, this plant is often found in distinct colonies. The several stems are glabrous, erect, and up to two feet tall. The pale green leaves are numerous and are narrowed to the petiole-like base. The flowers are closely set in a compact spike. The yellow corolla is very irregular, being strongly bilabiate; the upper lip has two lobes and the lower one has three. The lower lip has an elevated orange palate which closes the throat of the corolla. The corolla is also strongly spurred at its base with an awl-shaped tube. The fruit, round to ovoid, splits open irregularly to emit the winged seeds. Flowering may occur from June to October.

Scrophulariaceae
(Figwort family)

Penstemon albidus   White penstemon (Plate 189)

The penstemons have been cultivated by flower gardeners for many years and a society has been established for persons interested in the culture of penstemons. Glen Viehmeyer was especially interested in this group while he worked at the North Platte Experiment Station.

There are several wild species in our state but only a few will be shown and discussed here.

This species is found frequently in the sandhills and in the western part of the state. It blooms from June to August. Usually a few short stems occur close together arising from a persistent base of the previous year. The somewhat hairy stems bear leaves with toothed edges. The inflorescence has crowded flowers and is usually very hairy.

Flowers are less than an inch long and have white corollas which are sometimes tinged with lavender. The tubular part of the corolla tapers evenly and gradually and is not greatly inflated. The

94

Plate 178   Lemon monarda
Monarda pectinata

Plate 177   Wild bergamot
Monarda fistulosa

Plate 180   Heal-all
Prunella vulgaris

Plate 179   Catnip
Nepeta cataria

Plate 181   Blue salvia
Salvia azurea

Plate 182   Wood sage
Teucrium canadense

Plate 183   Jimson weed
Datura stramonium

Plate 184   Ground-cherry
Physalis heterophylla

Plate 186   Water-hyssop
Bacopa rotundifolia

Plate 185   Buffalo-bur
Solanum rostratum

Plate 187   Slender gerardia
Agalinis tenuifolia

Plate 188   Butter-and-eggs
Linaria vulgaris

Plate 189   White penstemon
Penstemon albidus

Plate 190   Narrow-leaved penstemon
Penstemon angustifolius

Plate 191   Cobea penstemon
Penstemon cobaea

Plate 192   Slender penstemon
Penstemon gracilis

Plate 193   Shell-leaf penstemon
Penstemon grandiflorus

Plate 194   Mullein
Verbascum thapsus

Plate 195   Water speedwell
Veronica anagallis-aquatica

Plate 196   Unicorn-plant
Proboscidea louisiana

Plate 197   Cancer-root
Orobanche fasciculata

Plate 198   Northern bedstraw
Galium boreale

five lobes which turn back abruptly are almost equal in size but distinctly form an upper lip of two segments and a lower lip of three. A close look at the corolla with a hand lens will show the capitate, glandular hairs which add sparkle to the appearance of the flowers.

Scrophulariaceae
(Figwort family)

Penstemon angustifolius   Narrow-leaved penstemon (Plate 190)

This plant is at home in western Nebraska in the sandhills, but it may be found as far east as Neligh. This photograph was taken near Taylor.

This plant, as indicated by its name, has long, narrow leaves. The stem, which becomes about a foot tall, is glabrous and the leaves lack marginal teeth. The bracts which subtend the flowers in the dense inflorescence are longer than the flowers in the lower part but shorter than the flowers in the upper part. The corolla tubes are narrow at the base but flare out to two or three times that diameter just below the middle of the tube. There are five lobes of the corolla in two lips. As usual, there are four fertile stamens. In this species the sterile stamen is slightly curved and hairy at the tip. The calyx is segmented into five narrow, nearly equal segments.

Scrophulariaceae
(Figwort family)

Penstemon cobaea   Cobea penstemon (Plate 191)

Cobea penstemon is one of two species of penstemon which have large flowers up to two inches long. The stems are stout and the leaves are wide with distinctly serrate margins. The leaves are a light green color and rather thick, and they tend to be clasping at the base.

The color variation in the flowers is very great. In the picture is the whitest form I have seen, but some variants have purple flowers similar to *Penstemon grandiflorus*. This plant is minutely hairy, attains a height of about two feet, and is found growing in dry, well-drained soil in southeastern Nebraska.

In average years the best blooms will be found in June.

Scrophulariaceae
(Figwort family)

Penstemon gracilis   Slender penstemon (Plate 192)

This penstemon is found frequently in northwestern Nebraska. The picture was taken during the last week of June in Chadron State Park.

The plants are somewhat clothed, especially the upper parts, with reflexed hairs in two lines down the stem. The narrow, waxy leaves are slightly toothed and are in pairs on the stem.

The inflorescence is diffuse and slender. Less than one inch long, the flowers are borne on short, lateral erect branches, and as shown here they are darker in color than commonly found. The plants are usually not much over one foot tall.

Scrophulariaceae
(Figwort family)

Penstemon grandiflorus   Shell-leaf penstemon (Plate 193)

This species is the other penstemon with large flowers, up to two inches long, that is common in Nebraska.

These plants are the largest of the native penstemons growing wild in the state. Under favorable conditions most plants which flower will be from two to four feet in height. The plants are without hairs on the stems and leaves, but these parts are usually glaucous. The thick leaves are broadly ovate to rounded and somewhat clasping at the base. The leaf margins are entire, not toothed or incised. The bract leaves shown in this picture are smaller than the flowers and smaller than the lower foliage leaves below the inflorescence.

The flowers of this species usually are light purple with open tubular corollas, as seen in this photograph. Flowering occurs mostly during June with plants found in dry sandy soil over most of the state.

Scrophulariaceae
(Figwort family)
Verbascum thapsus   Mullein (Plate 194)

One of our tall, erect, single-stemmed biennials, this plant is common in eastern Nebraska in disturbed areas during June, July, and August. Roadside specimens which have been mowed repeatedly may flower as late as September. During their first year of growth a rosette of grayish, very hairy leaves is produced. During the second year a tall flowering stem appears. The lower leaves are oblong to oblanceolate and petioled and up to a foot long. The upper leaves gradually diminish in size, become sessile, and also run down along the stem as a ridge of tissue.

The spike-like inflorescence is very densely covered with closely set yellow flowers which individually are not very exciting. The corolla is divided into five parts and the five stamens within usually all produce pollen. The calyx is heavily clothed with branched hairs as are the stem and leaves, and it is deeply segmented into five pointed segments of equal size.

Scrophulariaceae
(Figwort family)
Veronica anagallis-aquatica   Water speedwell (Plate 195)

This native of Europe has become well established along irrigation canals and slowly moving streams. It is sometimes found in shallow water, the shoots rising above the water (emersed) rather than floating on it. The stems and leaves arc essentially glabrous and the leaves are opposite, elliptic, sessile, clasping, and finely serrate.

The inflorescences are axillary and many-flowered. Each flower has a green calyx which is decply segmented. The corolla tube is very short and its divided segments are blue with deeply colored veins. The upper segment or petal has more veins than the others. This is one of the reasons taxonomists suggest that this represents a fusion of two segments of the corolla and thus the upper lip of an irregular flower. The flower is also unusual in having only two stamens. The single style with its single stigma is also prominent in the center of the flower. Flowering may occur throughout the growing season.

97

A terrestrial weed of lawns and cultivated areas, *V. agrestris* blooms in March and April from low, creeping stems bearing rotund, crenate, nearly sessile leaves.

Martyniaceae
(Unicorn-plant family)
Proboscidea louisiana   Unicorn-plant (Plate 196)

This coarse, strong-scented, clammy, diffusely branching annual bears simple rounded leaves which are cordate and unequal at the base. The lower leaves are opposite but the upper leaves are alternate. The plant may be two or three feet tall. Flowering usually occurs in July but occasionally may continue into September. This is not a common plant in Nebraska but may be found in fields and waste areas. The large flowers, up to two inches wide, occur in a short terminal raceme with dull white to yellowish tubular corollas having five nearly equal lobes which spread or reflex, leaving an open throat. Four stamens are included in the tube which is lined with purplish to brown spots.

Most striking of all the features is the fruit, which is a two-valved capsule with a long incurved beak. As the fleshy part of the fruit falls away the woody inner part remains. When the two valves separate, the long unicorn beak also splits. The fruit may reach a length of four to eight inches.

Orobanchaceae
(Broom-rape family)
Orobanche fasciculata   Cancer-root (Plate 197)

The common name is derived from the fact that many of the plants of this family have been found to grow parasitically on the roots of other plants. These herbaceous plants lack chlorophyll and thus must obtain their food from other plants. Their leaves are reduced to scales and the whole plant has a brownish to purplish color. A whitish glandular-pubescence may be seen on close examination. This picture was taken near the end of June on the top of Scotts Bluff National Monument in western Nebraska. This is quite a contrast to the moist woods which are usually listed as the normal habitat of *O. uniflora*, which is similar in appearance.

As seen in the picture the stem or caudex is mostly below ground and the elongate pedicels each support a single flower. The calyx is short-tubular with usually shorter teeth. The corolla is long-tubular with several longitudinal ridges. It is bent downward and the petal segments are small and oblong.

Rubiaceae
(Madder family)
Galium boreale   Northern bedstraw (Plate 198)

Most of the bedstraws that thrive in Nebraska are scrambling herbs whose few and dispersed flowers hardly admit them to a book of wild flowers. However, northern bedstraw, *Galium boreale,* is different in many respects. It is a perennial, found in northwestern Nebraska, which flowers from May to August. The stems are erect and not retrorsely barbed as is our common *Galium aparine.* The leaves are in whorls of four and vary from lanceolate to linear in shape. The leaves are three-veined and smooth with minutely rounded tips instead of sharp points.

The inflorescence is a somewhat dense terminal panicle bearing bright white flowers each of which is borne on a pedicel. The fruits which result from pollination and fertilization of the flower may be bristly or smooth. This character is the basis of several varieties that are commonly recognized by systematists.

Rubiaceae
(Madder family)
Houstonia nigricans   Narrow-leaved bluets (Plate 199)

Tufts of stems arise from a perennial base to a height of from six inches to a foot or more. The leaves are narrowly linear and sessile with but a single rib or vein. Short axillary fascicles are produced at many nodes.

Flowers are short-pedicelled and very numerous, forming crowded panicles. The calyx is composed of four narrow scabrous lobes. The corolla is tubular with widely spreading segments which are shorter than the length of the tube. These petals or corolla segments are four in number and are hairy on the inner (or upper)

99

surface. The color of the corolla is from purple to white with many gradations of color in between. Those with purple flowers often have strikingly bright blue anthers on the exserted stamens. In these flowers the throat of the corolla is wider and the style and stigma is included. In the picture, the other kind of flower in this dimorphic species is shown. Here the style and stigma are long-exserted and the anthers of the stamen come just to the throat of the tube which gives the impression of a narrow or closed throat on the corolla tube.

These plants grow in dry soil and flower during the months of May, June, and July.

Caprifoliaceae
(Honeysuckle family)
Triosteum perfoliatum   Horse-gentian (Plate 200)

These coarse, erect, hairy perennials with large perfoliate leaves are an eastern species reaching into eastern Nebraska where they are found in low areas having moist soil or on the moist bases of wooded hills. They flower from May through July but have persistent fruit which does not change to its characteristic color until late September or October.

The opposite leaves are obovate and at least some are joined at their bases around the stem.

The flowers are produced in small axillary clusters. The sepals are long and narrow and uniformly pubescent on the backs and on the margins. The corollas are hairy and purplish to dull greenish yellow.

The fruit is a dry berry which is crowned by the persistent sepals. In this variety the fruits are orange yellow.

Cucurbitaceae
(Gourd family)
Echinocystis lobata   Wild mock cucumber (Plate 201)

In the vegetative condition this high-climbing annual vine is characterized by long-petioled leaves with, normally, five sharp triangular lobes. It is found on alluvial soil, and it flowers during

100

August and September. In moist ditches and river bottoms it scrambles over other vegetation by the aid of three-forked tendrils.

This plant is monoecious, which indicates that each plant bears two kinds of unisexual flowers. The most conspicuous flowers are staminate, or pollen-producing. They are borne in long racemes as shown in the photograph of this species. The pistillate flowers have short peduncles and occur either solitarily or in a small cluster. The pistils have a two-celled ovary, in each cell of which two ovules are produced. When mature, the ovary is a bladdery, inflated, weakly prickly fruit about two inches long which dehisces by two terminal pores.

The whorls of flower parts each have six members. The calyx is very small; the corolla is white with six long white petal segments. The stamens are united by their filaments into a column. The style has a broad-lobed stigma. These last two parts appear in separate flowers.

Campanulaceae
(Bellflower family)

Campanula americana   Tall bellflower (Plate 202)

Tall bellflower is an eastern species which makes its home in woods and moist thickets along streams as far west as the eastern parts of Nebraska and Kansas. July is the month of fullest blooming, although flowering may continue into September. This herbaceous annual may attain a height of six feet. The leaves are thin and hairy, lanceolate in shape, and three to six inches long. The leaf lamina continues down the leaf to produce a margined or winged petiole. The margins of the leaf are serrate and the tips acuminate.

The upper part of the stem is an inflorescence up to a foot or more long. The lower bracts are leaf-like, but the upper ones are reduced. The flowers are solitary or in small tight clusters on short pedicels. The sepals are linear and spreading. The corolla is rotate, blue, and about one inch in diameter. Five stamens are attached at the base of the corolla by broad membranous filaments. The style is down-curved and then up-curved near its tip where it is divided into three or four stigmas.

101

Campanulaceae
(Bellflower family)
Campanula rotundifolia　　　　　　　　Harebell, bluebell (Plate 203)

These delicate plants grow in moist places and woods in western Nebraska and bloom from June to September. They vary in height and in shape and size of leaves and flowers. The picture was taken in Chadron State Park on June 27. This plant was over a foot tall and its stem leaves were linear. The shape of the flowers is campanulate (bell-shaped) and this is the basis of both the genus name *(Campanula)* and the family name (Campanulaceae). The lobes of the corolla are much shorter than the bell. The flowers are typically nodding in age.

Campanulaceae
(Bellflower family)
Triodanis leptocarpa　　　　Western Venus' looking-glass (Plate 204)

This annual species often grows among the grasses of our tall-grass prairies in eastern Nebraska. It may be found in bloom from May to August but is seen in most abundant flowering in late June and early July. The stems, which are often not over a foot tall, may bear flowers from near the base to the tip. The sessile leaves are lanceolate to linear and somewhat toothed. The flowers are sessile or nearly so with one or more occurring in the axil of each leaf. Often the lower flowers on the stem fail to open. Such non-opening flowers are said to be cleistogamous. In normal flowers the lobes of the violet corolla are longer than the tube. The calyx is often divided into four or five slender segments. As one observes in other bellflowers the stamens are free with the short filaments expanded at the base.

Campanulaceae
(Bellflower family)
Triodanis perfoliata　　　　　　　　Venus' looking-glass (Plate 205)

These annual plants may be found in woods, beside streams and roadsides, and in disturbed areas over most of the state. They may

102

blossom from May to September but in normal seasons they finish blooming by the end of June. The recognized genus name of these plants was formerly *Specularia.*

The stem is somewhat scabrous or hairy near the base. The rounded and usually toothed leaves are sessile and clasp the stem.

Flowers may occur singly or with a few at a node. The early basal flowers are usually small and cleistogamous, but the upper, later flowers have a lavender or purple corolla up to one-half inch long.

Lobeliaceae
(Lobelia family)

Lobelia siphilitica   Great lobelia (Plate 206)

The great lobelia was once thought to be a cure for syphilis— thus the specific name. These unbranched, stiff, erect, coarse perennials may attain a height of three feet. They are terminated by an open to dense leafy raceme. The plants vary from smooth to sparsely hairy, and the leaves are elliptic to lanceolate and sharply and irregularly toothed. The upper leaves become progressively smaller. Found along banks, shores, and other moist locations over most of the state, they are especially abundant in the sandhill swales and wet meadows. Blossoming occurs from July to October.

The two-lipped flowers are borne on pedicels with small bracts just above the middle. The corolla is one inch or more long and colored violet to deep rich blue, with white on the lower lip and into the throat. Some forms may be found with pink or even white corollas. The corolla tube is straight and is split down the middle on the upper side. There are two erect segments composing the upper lip, and three spreading segments in the lower lip. The five stamens, united into a tube by their filaments, enclose the slender style which protrudes through the upper corolla cleft and bends downward to mature in the sinus between the two segments of the upper lip.

Lobeliaceae
(Lobelia family)

Lobelia spicata   Pale-spike lobelia (Plate 207)

This species of lobelia is variable, with several named varieties. As shown here the flowers are white, but they are often described as pale blue. The unbranched stems, densely pubescent near the base and smooth at the tip, are not much over one foot tall. Flowering occurs from June to August. In the sandhills this plant grows in moist sand in meadows, fields, and thickets around lakes.

The leaves are oblanceolate, two or more inches long, and ascending along the stem.

The flowers are borne on pedicels with a pair of bracts at or near the base. The flowers are up to one-half inch long and have much the same form as in the preceding species.

Compositae
(Composite family)
Tribe Heliantheae

Bidens aristosa   Tickseed sunflower (Plate 208)

This species is commonly found in wet areas or on low ground throughout most of Nebraska. The size of this plant varies with its habitat. The specimen shown is relatively small (at two feet tall) since some plants exceed four feet in height. The internodes are long, displaying the leaves some distance apart and producing a diffuse plant which appears to have few leaves. The pinnately divided leaves have slender and irregularly toothed segments.

The heads are borne on elongated peduncles and have widespreading golden yellow rays up to one inch long. The flat fruits develop appressed hairs and a ciliate margin near the end, which gives rise to another common name, bearded beggar-ticks. There are two or four retrorsely barbed awns at the summit of the fruit. Flowering occurs from August into October.

104

Compositae
(Composite family)
Tribe Heliantheae

Bidens cernua Beggar-ticks, large bur-marigold (Plate 209)

This large bur-marigold may be found in swamps and wet meadows, on pond margins and in areas covered with water. It occurs frequently in Nebraska and flowers from August to November.

The stems and leaves are glabrous. The stems may grow to three feet in height and bear simple, opposite, lanceolate leaves which have serrate margins and may be perfoliate around the stem.

The heads are hemispheric, usually having eight ray flowers which are about one-half inch long. Each ray flower is yellow with lighter yellow (or white) tips. Drooping of the heads in age is the characteristic that gives the specific name of this plant.

The common name for the genus, beggar-ticks, comes from the fruit, which is dark, flat, and tipped with two or four spines that are retrorsely barbed so that they adhere to clothing very tenaciously.

Compositae
(Composite family)
Tribe Heliantheae

Coreopsis tinctoria Plains coreopsis, tickseed (Plate 210)

In central Nebraska during July, coreopsis may fill large areas of moist soil with its bright yellow flowers. Occasional small patches appear in eastern Nebraska in moist ditches but they rarely invade large areas. Flowering has been recorded as early as May and as late as September.

These annual plants grow to four feet, or even higher, in dense stands. The plants are devoid of hair and their leaves are opposite and divided pinnately into narrow segments.

The flower heads are borne on slender stalks and have large ray flowers which are three-lobed. Rays vary in number from five to seven and often show an uneven spacing around the head. The bases of the ray flowers may be a red or red brown color similar to, or brighter than, the color of the disk flowers in the center of the head. This serves as a distinctive and easily recognized characteristic. This plant is attractive and has been cultivated in flower gardens under the name calliopsis.

Compositae
(Composite family)
Tribe Heliantheae

Echinacea pallida   Purple coneflower (Plate 211)

As this species grows in Nebraska it is probably the variety *angustifolia* which in the past has been segregated as a separate species by some authors. On dry upland locations this plant is at home over most of the state. The photograph was taken on July 1 in the Dalbey prairie about five miles south of Virginia in Gage County. Most plants flower during June and July.

These plants are rough, hairy perennials usually under two feet in height. Their alternately arranged leaves, which are long and narrow with three main veins, occur near the base of the plant.

The flower heads, borne singly on noticeably long stems, are surrounded by pale purple or rose-colored ray flowers which droop at maturity. The disk flowers are on a dome-shaped base, making a ball-shaped central disk. The sharp tips of the bracts (chaff) of the disk flowers are longer than the disk flowers themselves and thus they produce a roughness that gives the genus name *Echinacea,* from the Greek, to this plant.

Compositae
(Composite family)
Tribe Heliantheae

Helianthus annuus   Common sunflower (Plate 212)

The cultivated sunflower, with the enormous head which produces the sunflower seed of commerce, belongs to this species. The wild forms have smaller heads but can still be considered large. The plants usually exceed four feet in height and have rough, branched stems. As indicated by the specific name, the plants are annuals, flowering from July to September.

Leaves usually occur singly on the stem, but occasionally they may arise in pairs near the base of the stem. Ordinarily, the leaves are supported by distinct petioles and the blades are broad and almost deltoid with the leaf base being somewhat cordate, the margins toothed, and the leaf surface rough.

Flower heads vary greatly in size. The ray flowers overlap one another at least part of the way around the head and the disk flowers produce a brown center. All species of the genus *Helianthus* have nonfertile ray flowers. The bracts on the underside of the flower head are broad, edged with bristles, and abruptly narrowed to a point. This species is often difficult to distinguish from *Helianthus petiolaris.*

Compositae
(Composite family)
Tribe Heliantheae
Helianthus grosseserratus   Saw-toothed sunflower (Plate 213)

This is one of our tallest sunflowers. It grows in damp prairies, bottom lands, and other moist places. The leaves are lance-shaped with coarse teeth and short, often winged, petioles. Except for the lowest ones, the leaves are attached alternately and often show a whitish pubescence on the lower surface. The upper surfaces of the leaves are not rough or only scarcely so. The stems are strigose near the flower heads but are without hairs below, becoming smooth and often glaucous.

The flower heads may be few or several on the upper branches. They appear from July to October. The ten to twenty yellow ray flowers surround the yellow disk which may be up to one inch in diameter.

Compositae
(Composite family)
Tribe Heliantheae
Helianthus rigidus   Stiff sunflower (Plate 214)

A previously used scientific name for this species was *Helianthus laetiflorus* var. *rigidus.* The common name, stiff sunflower, remains fairly descriptive of these small perennial plants which produce slender, mostly unbranched stems with stiff leaves. The lower leaves are paired and close together, but the upper leaves are more distant and may occasionally occur singly at the upper nodes. The leaves have lanceolate blades which taper into very short petioles.

The flower heads are borne on long, slender, usually rough terminal stems or on stems which arise from the axil of an upper leaf. The bracts subtending the flowering head are broad, firm, appressed, and overlapping. The fifteen or more yellow ray flowers surround a disk which is usually dark, but in some variants it may be yellowish. Flowering occurs in August and September.

Compositae
(Composite family)
Tribe Heliantheae

Helianthus maximiliani   Maximilian's sunflower (Plate 215)

These perennial plants are usually found clustered together as indicated in the picture. The stems attain a height of three to five feet when growing in their typical roadside, prairie, or sandy waste place environments.

One of the easiest means of identifying this species is by its leaves. Most of the leaf blades are partially folded upward along the midrib like a keel and the midrib is arched downward (falcate). The narrow, alternate leaves are very rough on both the upper and the lower surfaces, and they taper into a short, winged petiole.

The ray flowers often curve upward around the yellow disk. The bracts of the head are narrow, loose, and often longer than the disk. The picture was taken in late September; flowering of this species has been observed from July to October.

Compositae
(Composite family)
Tribe Heliantheae

Helianthus petiolaris   Prairie sunflower (Plate 216)

Prairie sunflowers are often confused with common sunflowers. The prairie sunflower is a slenderer plant whose leaves are somewhat narrower in proportion to their length, and the leaf margins are less likely to have teeth as pronounced as those of the common sunflower. The bracts on the back of the flower head are lance-shaped, tapering gradually to the tip. The edges of the bracts are not lined by long bristles as in the common sunflower although there may be a few short bristles along the margin.

During midsummer before most sunflowers are in bloom, the prairie sunflowers, which are found in dry sandy soil, may be in flower though scarcely two feet tall. This does not mean that all short early-blooming sunflowers are of this species nor does it mean that this species may not grow tall. Compared to the common sunflower the prairie sunflower branches less and the branches are more ascending than spreading. Flowering may occur from June to September.

There are fifteen to twenty yellow ray flowers surrounding the brownish to dark purple disk. As in all sunflowers the disk flowers are fertile, but the ray flowers are not.

Compositae
(Composite family)
Tribe Heliantheae

Helianthus tuberosus   Jerusalem artichoke (Plate 217)

People who study the derivation of names tell us that the common name had nothing to do with the Holy Land but comes instead from a corruption of an Italian word which means turning toward the sun (Rickett, 1966).

One of the tall sunflowers, it has stout stems up to ten feet or more in height. Tubers are present on the underground parts fairly near the base of the stems. The stems are more or less hairy below the flowering heads. The lower leaves are opposite but the upper leaves are alternate. At least one-third as wide as long, they vary in shape from broadly lanceolate to ovate. Winged petioles, almost four inches long, support the thick leaf blades which are rough and have three main veins at the base.

The light yellow ray flowers are very conspicuous compared to the size of the small disk of yellow flowers. This species commonly flowers in September.

109

Compositae
(Composite family)
Tribe Heliantheae

Heliopsis helianthoides   Ox-eye (Plate 218)

The ox-eye, often called the false sunflower, is commonly found in eastern and central Nebraska and is to be found generally in the central and eastern United States. It is a perennial herb with opposite leaves borne on distinct petioles. The leaf base is often truncate. The leaf margins are coarsely toothed and the leaf surface rough. When they grow along roadsides, these plants attain a height of three to five feet. Three or more flowering heads are produced on smooth stalks from the main stem or from the axils of the upper leaves.

Each inflorescence has more than ten yellow ray flowers surrounding a conical head of yellow disk flowers. The ray flowers produce a fertile pistil which will bear a fruit (incorrectly called a seed), as do the disk flowers. In the sunflowers *(Helianthus)* the ray flowers are not fertile. Flowering of plants which are growing in full sunshine usually occurs in July but may continue until September.

Compositae
(Composite family)
Tribe Heliantheae

Ratibida pinnata   Gray-headed prairie cone-flower (Plate 219)

Cone-flowers are easily recognized by the tall receptacles or cones on which the disk flowers are borne. Of the two species of the genus *Ratibida* that are common in Nebraska, *R. pinnata* is the less common one. This species is a little taller, growing to four or five feet in height, and is found most frequently in eastern Nebraska. Flowering most commonly occurs in July and August. The preferred habitat is in low moist pastures, along streams, and in ravines.

The branched stems bear pinnately divided leaves which have from three to seven lanceolate segments.

The flower heads are terminal and erect on long slender stalks. The ray flowers, which are an inch or more long, droop down

110

Plate 199   Narrow-leaved bluets
Houstonia nigricans

Plate 200   Horse-gentian
Triosteum perfoliatum

Plate 201   Wild mock cucumber
Echinocystis lobata

Plate 202   Tall bellflower
Campanula americana

Plate 203   Harebell
Campanula rotundifolia

Plate 204   Western Venus'
looking-glass
Triodanis leptocarpa

Plate 205   Venus' looking-glass
Triodanis perfoliata

Plate 206   Great lobelia
Lobelia siphilitica

Plate 207   Pale-spike lobelia
Lobelia spicata

Plate 208   Tickseed sunflower
Bidens aristosa

Plate 209   Beggar-ticks
Bidens cernua

Plate 210　　Plains coreopsis
Coreopsis tinctoria

Plate 211　　Purple coneflower
Echinacea pallida

Plate 212　　Common sunflower
Helianthus annuus

Plate 213　　Saw-toothed sunflower
Helianthus grosseserratus

Plate 214　　Stiff sunflower
Helianthus rigidus

Plate 215   Maximilian's sunflower
Helianthus maximiliani

Plate 216   Prairie sunflower
Helianthus petiolaris

Plate 217   Jerusalem artichoke
Helianthus tuberosus

Plate 218   Ox-eye
Heliopsis helianthoides

Plate 219   Gray-headed prairie
coneflower
Ratibida pinnata

around the supporting stalk. The disk is spherical to slightly elongate, but it is usually never more than twice as tall as wide, and it is shorter than the ray flowers. After the blooming of the disk flowers, the disk is dark brown instead of grayish.

The more common species, *Ratibida columnaris,* is found in dry prairies throughout the state. Its cones may be four times as tall as wide and are almost invariably cylindrical. The cone is usually longer than the drooping ray flowers. The sequence of flowering of the disk flowers from the base to the apex of the cone can easily be discerned. Its common name is prairie cone-flower.

Compositae
(Composite family)
Tribe Heliantheae

Rudbeckia hirta   Black-eyed Susan (Plate 220)

This species has also been called brown-eyed Susan and some authorities recognize a different species as black-eyed Susan. This one is widely distributed in the prairies and plains regions of the state and is becoming a common weed in the eastern part of the state. It may be in flower during any of the summer months.

The stems are less than three feet tall with limited branching and terminal flower heads. The leaves are not lobed and their margins vary from entire to toothed. The plants are usually rough and hairy in all parts but the degree of hairiness may vary considerably.

The flower heads have yellow to orange ray flowers which are usually conspicuous. The disk is more or less spherical and dark brown at maturity.

The double-flowered cultivated form called golden glow belongs to another species *(R. laciniata)* of this genus.

Compositae
(Composite family)
Tribe Heliantheae

Silphium integrifolium   Entire-leaved rosinweed (Plate 221)

Three species of the genus *Silphium* are commonly seen in flower in eastern Nebraska during July and August. This species is perhaps the most common of the three. As the species name indi-

111

cates, the leaf margins are not cut up, being entire or only slightly toothed. The leaves are thick and rough on both surfaces and are sessile and set in pairs upon a cylindrical stem. The height of the plant at full bloom may be up to five feet. There are several flower heads on each stem. The bracts of the head in all species are similar. They are loose and leaf-like. The ray flowers are yellow and fertile, and the disk is also composed of yellow flowers but all of them are sterile and they bear no achenes.

Silphium laciniatum, the compass plant, has large dissected leaves on tall stems. The leaves stand mostly erect and are supposed to be predominantly oriented with their edges north and south.

Silphium perfoliatum, the cup plant, has opposite leaves that join together at their bases around the square stem, forming a cup that will hold rain water at each node.

Compositae
(Composite family)
Tribe Heliantheae

Thelesperma filifolium   Nippleweed (Plate 222)
var. *intermedium*

These small annual or biennial plants are at home in the sandy soil of the western part of the state. For the most part, they may be not much more than a foot in height, although in favorable growth situations they have been reported to be up to three feet tall. The stems are branched throughout and they bear many flower heads. The leaves are divided into narrow linear segments which are not rigid as in other species. The rays, large and three-lobed at the tip, surround the yellow disk flowers. Flowering has been reported from June to August.

Thelesperma gracile, the rayless thelesperma, has heads on long, slender stalks that never seem to stop their movement even in unnoticeable air currents. The flower heads have ray flowers that are inconspicuous or may be lacking. This is a feature which makes this species seem to be an unlikely member of this tribe.

Compositae
(Composite family)
Tribe Helenieae

Helenium autumnale   Sneezeweed (Plate 223)

An inhabitant of swamps, stream banks, and wet places, the sneezeweed has established itself throughout most of the United States. This is a perennial which grows to three feet or more in height and flowers in late summer or autumn.

The leaves are alternate on the stem, coarsely dentate, and usually decurrent, making the stem somewhat winged. The stem may be unbranched or, if it is branched near the summit, each branch bears a flower head.

Ray flowers surround the globe-shaped disk more or less evenly and are about as long as the disk is wide. They are broadly wedge-shaped, usually with three rounded lobes at the outer end.

Compositae
(Composite family)
Tribe Helenieae

Hymenoxys acaulis   Stemless hymenoxys (Plate 224)

This unobtrusive, low-growing plant is often overlooked in our dry, sandy prairies and hillsides except in June and July when it comes into full bloom. Winter (1936) suggests that flowering is from May to August.

The solitary yellow heads terminate the leafless flower stalk which arises from a cluster of punctate basal leaves to a height of one foot. The wedge-shaped ray flower corollas terminate in three small rounded or obtuse lobes. There are usually more than ten ray flowers which are pistillate, fertile, and only occasionally overlapped laterally. These surround a center of yellow, perfect, fertile flowers. The erect involucral bracts are in two sets and have rounded tips.

The photograph of a plant growing in partial shade was taken at the Cochrane Wayside south of Crawford, Nebraska.

113

Compositae
(Composite family)
Tribe Anthemideae

Achillea lanulosa   Yarrow (Plate 225)

The low, flat, white-topped inflorescence supported on stems bearing minutely dissected leaves portrays the concept of yarrow to most nature lovers. Those in the illustration, having pinkish flowers, shows that color variation occurs. This species is a relatively common weed which has been introduced from Europe and now is found over much of the United States and is common in eastern Nebraska.

This rather strong-scented perennial is commonly under two feet tall. The plant is densely woolly and often unbranched up to the inflorescence. Each small head produces only a few small ray flowers, causing it to resemble a five-petaled flower. There are three or four series of involucral bracts subtending the flowers in each head.

Yarrow is found in flower in dry disturbed areas and in many of our prairies from June to October.

Compositae
(Composite family)
Tribe Anthemideae

Chrysanthemum leucanthemum   Ox-eye daisy (Plate 226)

Ox-eye daisy is one flower known to nearly everyone. It was introduced from Europe and is still grown in gardens, but it has also become well established in fields, in pastures, in waste places, and along roadsides in Nebraska and much of the United States.

The familiar white-flowered daisy has ray flowers often called "petals" which are white and disk flowers which are yellow. The whole head may be up to two inches across. The yellow disk has a dimple-like depression in its center. Close observation will reveal that the outer disk flowers mature first and elongate somewhat. As the disk flowers mature toward the center the depression becomes smaller. Flowering may continue from May to August.

The stems are erect and sparingly branched toward the top; they attain a height of one to three feet. The basal leaves have

114

long, slender petioles while the middle and upper leaves are oblong with spreading teeth near their junction with the stem. All leaves are coarsely but regularly toothed.

Compositae
(Composite family)
Tribe Senecioneae

Cacalia tuberosa   Tuberous Indian plantain (Plate 227)

From June to August this stout perennial may be in flower in wet swampy areas and low moist prairies of the eastern part of Nebraska. The angled stems bear whitish flowering heads at their tips, three to four feet above the soil. The leaves are green on both surfaces and are striately nerved, with five or more strong vascular bundles traversing the leaf and converging near the leaf tip. The leaf is very similar to a plantain leaf because the margin is essentially entire and its base tapers into a long petiole.

The white flower heads are composed of five tubular flowers which bear both stamens and pistil. These are surrounded by five main involucral bracts. The corolla is deeply five-cleft.

The fruits are oblong and smooth and are surmounted by numerous soft hair-like bristles.

Compositae
(Composite family)
Tribe Senecioneae

Senecio plattensis   Ragwort, senecio (Plate 228)

This low-growing, yellow-flowered composite is one of the small early flowers of the prairies of Nebraska. These plants grow to heights of between one and two feet and bear a single head or, more usually, a small cluster of heads at the tip of the stem. The leaves are alternate in their arrangement on the stem with the basal ones sometimes forming a rosette. Basal and lower leaves are usually unlobed at their bases, while the upper leaves are smaller and many-lobed and may be covered with fine woolly hairs on the lower surface.

115

The heads are showy with conspicuous yellow ray flowers surrounding a yellow disk. The involucral bracts are narrow and slender-tipped. The bare receptacle, fringed with dry bracts, persists well into the middle of the summer in many of our prairies.

Compositae
(Composite family)
Tribe Astereae
Aster ericoides   Heath aster (Plate 229)

A name often used for this plant by the students of the late Professor J. E. Weaver was *Aster multiflorus*, the many-flowered aster. This plant is common throughout the state and may be found in flower from August to November in roadsides, open prairies, and dry soil.

The plants usually attain a height of three feet or more by the time of flowering. They may be recognized in the prairie much sooner by their repeated branching and their many small, linear leaves.

The numerous heads are small and are fringed with about twenty white ray flowers. The anthers of the many disk flowers give a yellowish color to the whole disk.

Compositae
(Composite family)
Tribe Astereae
Aster novae-angliae   New England aster (Plate 230)

This species is one of our most beautiful asters. Its flowering heads are large and the ray flowers are longer than those of any other of our asters. These plants inhabit moist areas and display their bright flowers during September in many places in the eastern half of Nebraska. Water-filled ditches beside hay meadows are their favorite habitat. They have been found in flower from August to October.

The mature plants may attain a height of six feet. Their lanceolate leaves have entire margins and are about four inches long. The leaf is sessile with a lobed, somewhat clasping base that partly surrounds the stem. The flower heads have approximately forty ray flowers with purple corollas about three-fourths of an inch long. This plant is well deserving of a place in wild flower gardens.

Compositae
(Composite family)
Tribe Astereae

Aster sericeus   Silky aster (Plate 231)

The silky aster has perhaps a more distinctive vegetative growth habit than many other plants. During midsummer this aster may easily be identified by its stem and silky leaves. The stems are smooth, straight, and brittle. Its color is often a brownish red. The few, divergent branches of the stem arise in the upper half of the plant. The leaves are small (about an inch long), entire, and sericeous (covered with silky hairs that are appressed to the leaf surface). The lower leaves fall during the growing season, leaving only upper leaves at the time of flowering.

Several heads are usually clustered at the ends of branches. About twenty purple ray flowers surround the head and each is approximately one-half inch long. These plants are at home in the dry prairie areas and other open dry ground in the eastern part of Nebraska. Flowering occurs during August and September.

Compositae
(Composite family)
Tribe Astereae

Chrysopsis villosa   Prairie golden-aster (Plate 232)

One popular manual by Gleason and Cronquist (1963) lists five varieties of this species. In Nebraska various varieties may be recognized. The more hairy form, var. *foliosa,* tends to be found in western sandy soil while var. *villosa,* which is less hairy with long peduncles, is more eastern but shows considerable variation in other characteristics.

The plant is perennial with several decumbent stems arising from a stout rootstock. Depending on soil and environmental conditions the tips of the branches may rise from one-half foot to al-

117

most two feet above the soil. The leaves are entire and mostly sessile but some upper leaves may have short petioles.

The numerous heads are bright yellow because of the color of the ray flowers. The disk flowers are yellow also and become brownish as they age. If the fruit is observed carefully the pappus may be seen to be double. The inner pappus is of thread-like bristles while the outer pappus is of scales. The fruits of both the ray flowers and the disk flowers are flattened. Flowering occurs during July and August.

Compositae
(Composite family)
Tribe Astereae

Erigeron pumilus   Low erigeron (Plate 233)

There are several low-growing erigerons in the state and this species is perhaps the most common one in the sandhills and in northwestern Nebraska. The heads are numerous and showy but the disk is usually less than one-half inch in width. The stems are tufted, bearing long bristly white hairs. The leaves are mostly linear. The ray flowers are usually white but variations to blue may occur. Flowering may extend from May to September.

Compositae
(Composite family)
Tribe Astereae

Erigeron strigosus   Daisy fleabane (Plate 234)

Usually annual plants with few leaves, these white-flowered plants are at home in prairies, rocky open places, dry fields, and open woods. They are considered weeds over most of the United States. The basal leaves are oblong and toothed or entire, while the upper leaves are usually entire.

The stems are branched, producing several to many heads. The length of the white ray flowers is about equal to the width of the whole disk. The disk is composed of many yellow disk flowers.

This is probably our most common species, and it blossoms in greatest abundance during the summer months of June and July

118

although it may be seen sporadically through most of the growing season. Its normal height is about two feet.

Compositae
(Composite family)
Tribe Astereae

Grindelia squarrosa   Gumweed (Plate 235)

Once started in a pasture this sticky weed may take over large areas, especially when overgrazed conditions prevail. It grows to about two to two and one-half feet tall and is branched in the upper part. The lower leaves may have dropped off by the time flowering occurs. The leaves are more or less toothed and the upper leaves tend to clasp the stem.

The many upper branches each bear a sticky flower surrounded by many narrow bracts which are squarrose. This gives the species name to the plant. The heads are about one inch across when mature and both ray and disk flowers are yellow.

These plants grow throughout Nebraska on dry prairies and are in flower from late June to early September.

Compositae
(Composite family)
Tribe Astereae

Haplopappus spinulosus   Cut-leaved goldenweed (Plate 236)

Of frequent occurrence in western and west central Nebraska, this perennial grows with many stems arising from the rootstock. The young shoots and leaves are often somewhat woolly but become smooth with age. The stems, which may be up to two feet in height, bear very pinnatifid and bristle-tipped leaves.

The flower heads have distinctive yellow ray flowers and a set of yellow disk flowers in the center. The heads may be over an inch in diameter. They may be found in flower during any of our growing months but mainly during July.

Compositae
(Composite family)
Tribe Astereae

Machaeranthera sessiliflora   Viscid aster (Plate 237)

Often this species is included in the genus *Aster*. These biennial plants have leafy stems up to a foot and one-half tall. The leaves are linear with somewhat toothed margins; they are about one and one-half inches long and are slightly rough but not very hairy. The bracts of the head are awl-shaped, hairy, and only slightly glandular. The heads have conspicuous ray flowers which vary in color from rose purple to purple. This is a plant of the hills and plains and it flowers in August and September.

Usually the species *M. canescens* has ray flowers which are dark bluish purple and stem leaves which are linear and entire.

Compositae
(Composite family)
Tribe Astereae

Solidago gigantea   Giant goldenrod (Plate 238)
var. *serotina*

A specimen of this variety was taken to the Nebraska legislature by Professor Bessey at the time he proposed that the goldenrod be adopted as the state flower. Formerly this was *Solidago serotina* but recently systematic botanists have decided a more nearly correct name would be that given above.

This is a perennial plant which may grow as tall as six feet in favorable locations. The plant is minutely pubescent only in the inflorescence. The main stem is smooth and glaucous. The leaves of this variety are wholly glabrous on both sides; no hairs occur on the midrib or main veins as they do in the variety *gigantea*.

The branches of the inflorescence are covered on one side with small heads of several flowers each. The branches of the inflorescence are recurved.

The basal and lower stem leaves are smaller and they fall off before flowering occurs. The middle and upper stem leaves are larger (up to six inches long) and become only slightly shorter near the inflorescence. Flowering occurs from August to October.

Solidago nemoralis is a short, early-flowering goldenrod which may bloom in July yet continue into November. The height of the plant at the time of flowering is about two feet.

Compositae
(Composite family)
Tribe Astereae

Solidago rigida   Stiff goldenrod (Plate 239)

The leaves on this plant are rather stiff and harsh to the touch. The lower leaves have long petioles and oval or oblong leaf blades, while the upper leaves have no petioles; thus they are sessile and slightly clasping. Their shape is usually oblong also. The stems are likely to be branched only in the inflorescence at the top of the plant. In flower these plants may attain a height of three feet.

The inflorescence is often flat-topped and may be ten inches wide but is frequently not more than half that size. The heads making up the inflorescence are erect and showy containing up to forty flowers per head. The bracts below each head are broadly rounded at their ends.

These are frequently found in sandy soil throughout the state. Flowering occurs from August to October.

Compositae
(Composite family)
Tribe Astereae

Townsendia grandiflora   Townsendia (Plate 240)

In Nebraska most of these plants are found in Sioux County. They may flower as early as June and may continue until fall. The plants are perennial, producing several unbranched stems which bear a single flower head. Flowering usually occurs when the heads are not more than an inch or two above ground level. The flower heads often are surrounded by many leaves which are widest near the tip. Generally they are elongate and are attached singly to the stem.

121

The conspicuous ray flowers are longer than the disk is wide and are often colored a rose purple. The disk flowers are yellow and the width of the whole head is from one to two inches. This is unusual for flowers borne so close to the ground on such short stems.

Compositae
(Composite family)
Tribe Inuleae

Antennaria neglecta   Pussy-toes (Plate 241)

Although these plants are not very noticeable they are abundant in the state, growing in dry ground of pastures, prairies, and meadows and appearing before the warm-season grasses begin their annual growth. Some Rocky Mountain species may be somewhat more colorful and more conspicuous. There are several species in the state but their similarity of form and their variation within a clone or group make them a difficult group. Apomictic reproduction probably contributes to this variation and to the associated difficulties in classification.

The stoloniferous habit of these small plants, mostly less than six inches tall, causes them to grow in patches. The plants are dioecious, producing only staminate flowers on one plant and only pistillate flowers on another. Ordinarily all the plants of one patch will bear only one type of unisexual flowers. To find the other type of flower one must locate another patch. As in all composites each head is composed of many small flowers.

Compositae
(Composite family)
Tribe Eupatorieae

Eupatorium perfoliatum   Boneset (Plate 242)

A common sight on wet prairies over the state, this plant attains a height of three feet or more and flowers in August and September. The opposite, perfoliate leaves taper gradually from the base to the long acuminate tip. Leaf margins are crenate-serrate, and the leaf surface is rugose-reticulate. The leaves are usually densely pubescent on the under surface, and their midribs are usually

122

curved downward. The stem is conspicuously hairy with long, spreading hairs. Several axillary branches arise from the upper nodes.

The inflorescence is generally flat-topped with each head bearing several to many disk flowers having white corollas. The involucral bracts around each head are pointed at the tip and pubescent.

Other species of this genus which are often found in Nebraska are *E. maculatum,* joe-pye weed, with whorled leaves, *E. altissimum* with three-veined, sessile, opposite leaves having nearly entire margins, and *E. rugosum* with thin, broad, coarsely toothed leaves.

Compositae
(Composite family)
Tribe Eupatorieae

Kuhnia eupatorioides   False boneset (Plate 243)

One of the perennial forbs of the dry prairies and plains which can be identified during most of the summer from its vegetative growth is the false boneset. It grows to about two feet in height and bears alternate, sessile, lanceolate, sharply dentate leaves. The leaves, which are up to three inches long, are conspicuously veined.

The heads are numerous, erect, and one-half inch or more high. The outer bracts of the heads are narrow and slender-tipped. All of the flowers of the head are disk flowers and these are usually creamy white. The fruits have many ribs and the hairs at the summit are tan. Flowering occurs from August to October.

Authors who recognize several varieties of this species call our plants *Kuhnia eupatorioides* var. *corymbulosa.* These plants are stouter and more hairy and bear firmer leaves than other varieties such as *K. e.* var. *eupatorioides.*

Compositae
(Composite family)
Tribe Eupatorieae

Liatris aspera   Rough gay-feather (Plate 244)

This species like all others of the genus *Liatris* is perennial. Stems up to three feet in height arise from a corm less than an

123

inch thick. The leaves are numerous and rough, especially on the margins. The species name, *aspera,* which means rough, is the same as the first part of the common name. The lowest leaves may be as much as eight inches long while the upper leaves become smaller as they extend into the terminal inflorescence where they disappear from sight.

The flower heads bear many disk flowers. The corolla of a single flower when removed from the head and torn open will show a distinct hairiness at the base. The purple bracts surrounding the flowers are broadly rounded and loosely arranged. They are unevenly toothed on the inrolled dry edges. The heads, which tend to be sessile, form long but somewhat loose spikes. The bristles arising from the top of the fruit are not plumose but have minute barbs. Flowering may occur from August to October.

Compositae
(Composite family)
Tribe Eupatorieae

Liatris punctata   Dotted gay-feather (Plate 245)

Some caution must be used with common names. The names blazing star and gay-feather are used almost interchangeably with the species of this genus. Likewise, most of the species of this genus are punctate; that is, they have translucent dots in the leaves as seen with a hand lens when a leaf is held up to a bright sky.

Many stems arise from a single corm to a height of one and one-half to two feet. The leaves, which have one main vein, are alternate, strongly dotted, and hairy along the margin; but otherwise they are without hairs.

The numerous closely set heads are sessile on the stem and have only four to six flowers per head. The spike is leafy-bracted more or less throughout its length, which may be about six inches. The bracts beneath the flowers in the heads are hairy along the margins and are terminated by slender tips. The corollas are rose to purple and are hairless on the inside. The fruits have bristles which are plumose.

These plants are common on dry soil throughout the state and flower from August to October.

124

Compositae
(Composite family)
Tribe Eupatorieae

Liatris pycnostachya   Gay-feather (Plate 246)

These plants attain a height of five or more feet but may flower when only two feet tall if conditions are unfavorable. They grow in prairies throughout the state and flower from July to September.

The alternate, nearly linear leaves of this plant are numerous. The basal ones are especially long (up to a foot or more) and the upper ones short and narrow.

The flower heads are numerous, sessile or essentially so, and crowded on the stem, forming a spike. Each head has fewer than fifteen flowers. These few disk flowers are subtended by abruptly pointed bracts which are turned outward. If a single flower is picked from a head and the corolla torn open it should have no hairs on the inside of the corolla tube. The fruits have bristles which are strongly barbed but not plumose.

Compositae
(Composite family)
Tribe Eupatorieae

Liatris squarrosa var. *glabrata*   Blazing star (Plate 247)

This is a perennial herb which grows each year from a globose corm. The glabrous stems are longitudinally grooved and may grow occasionally (though not often) to a height of two feet. The leaves are alternate, without hairs, narrow, and two to six inches long.

The heads are few, sessile, and somewhat separated along the stem. The bracts of the head are callous-margined and hairy. The outer ones appear ovate in shape with short slender tips which are slightly spreading. The inner bracts are oblong and pointed. The heads have only disk flowers and their corollas are one-half inch long and red purple in color. The hairs on the fruits are plumose and slightly purple.

This plant is common throughout the state, growing on dry prairies, sandhills, and bluffs. Flowering occurs from late July to

125

September. Varieties of this species have been distinguished mainly by the degree of hairiness of the plant, particularly in the inflorescence.

Compositae
(Composite family)
Tribe Vernonieae

Vernonia fasciculata   Ironweed (Plate 248)

These plants typically do not grow in large clumps, but several stems may arise from a perennial rootstock. The tall, leafy, erect plants bear broad, relatively flat-topped, compound inflorescences made up of many heads with purple flowers. They are found in moist soil of marshy areas and low well-grazed pastures, yet they appear almost as specimen plants, virtually untouched by cattle. Their flowering season is from July to September.

The leaves are glabrous on both surfaces, as are the stems. The leaves are linear-lanceolate with slender tips and sharply serrated margins. The leaves, from four to eight inches long, are arranged alternately on the stem and have conspicuous pits on the lower surface.

The flower heads bear fifteen to thirty flowers that are surrounded by bracts which are slender, glabrous on the surface, and entire or slightly ciliate along the margin. The bracts have rounded tips and are erect. When mature, the fruits have two sets of bristles at the top. Short bristles or scales make up the outer circle while the inner circle consists of long bristles. This last characteristic is common to other species of this genus.

Compositae
(Composite family)
Tribe Cynareae

Carduus nutans   Musk thistle (Plate 249)

Since this species was introduced into the United States from Europe it has spread rapidly, and in 1959 it was added to the noxious weed list for Nebraska.

Unlike the other thistles, leaves of the musk thistle lack the white matted hairs on both upper and lower surfaces. They are

126

Plate 220   Black-eyed Susan
Rudbeckia hirta

Plate 221   Entire-leaved rosinweed
Silphium integrifolium

Plate 222   Nippleweed
Thelesperma filifolium
var. *intermedium*

Plate 223   Sneezeweed
Helenium autumnale

Plate 224   Stemless hymenoxys
Hymenoxys acaulis

Plate 226   Ox-eye daisy
Chrysanthemum leucanthemum

Plate 225   Yarrow
Achillea lanulosa

Plate 227   Tuberous Indian plantain
Cacalia tuberosa

Plate 229   Heath aster
Aster ericoides

Plate 228   Ragwort
Senecio plattensis

Plate 230   New England aster
Aster novae-angliae

Plate 231   Silky aster
Aster sericeus

Plate 232   Prairie golden-aster
Chrysopsis villosa

Plate 233   Low erigeron
Erigeron pumilus

Plate 234   Daisy fleabane
Erigeron strigosus

Plate 235   Gumweed
Grindelia squarrosa

Plate 236   Cut-leaved goldenweed
Haplopappus spinulosus

Plate 237   Viscid aster
Machaeranthera sessiliflora

Plate 238   Giant goldenrod
Solidago gigantea var. *serotina*

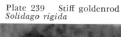

Plate 239   Stiff goldenrod
Solidago rigida

coarsely lobed, with three to five teeth per lobe. Each lobe ends in a spine. The edges of the leaf are wavy and the leaf blade seems to continue on down the stem, forming an interrupted, lobed, and spiny wing. The stems are variously hairy, ranging from nearly glabrous to a thin cobwebby condition but they never attain a white woolly condition. The stems often grow to three feet in height, branch freely, and often bear one or more inflorescences or heads in addition to the terminal one.

The heads are mostly nodding or somewhat so and the involucral bracts are somewhat sepal-like, the lower ones often strongly reflexed. The flower heads are reddish purple and roundish when in full bloom.

Flowering may start in June and continue until September or perhaps to October. As seen with a hand lens the pappus of the mature fruits are covered merely by minute barbs in contrast to the other thistles which have plumose pappus filaments.

Musk thistles are found in greatest numbers in southeast Nebraska, but they extend occasionally even to northwestern Nebraska.

Compositae
(Composite family)
Tribe Cynareae

Cirsium altissimum   Tall thistle (Plate 250)

The furrowed stems of the tall thistle may bear flowering heads at heights of from three to six feet. These plants are common over most of the state, growing either as biennials or perennials on non-cultivated land such as woodlands and hills or lowlands where soil drainage is poor.

The leaves are dark shiny green on the upper side and white woolly beneath. The basal leaves are over a foot long and deeply pinnatifid into narrow segments which themselves are lobed and spine-tipped. The upper leaves are much smaller with the uppermost ones fitting up closely to the terminal flowering head.

The flowering heads form in May or June, yet the plants fail to blossom until September or October. The many buds terminate the well-branched upper stems. These flowering stems are green

127

and slightly hairy. The flowering heads, about an inch in diameter, bear rose purple flowers. The bracts are numerous, the outer ones bearing a longitudinal ridge that continues as a weak, spreading, long tip that may be thought of as a colorless bristle.

Compositae
(Composite family)
Tribe Cynareae

Cirsium arvense   Canadian thistle (Plate 251)

New shoots of Canadian thistle arise in patches each year from an extensive underground root system. The leaves are oblong to long-pointed with medium to coarse lobes. The leaf margins are spiny and usually wavy. The leaves are dark green on the upper side and very light green and slightly hairy on the lower side, but some plants may have leaves which are smooth on both sides. The leaves are short-stalked, their bases often clasping the stem. The stems, slender compared to other thistles, are erect and hollow, vary from smooth to slightly hairy, and grow from two to four feet tall. The stems branch at the top and lack spiny wings.

The flowers of this plant are dioecious. The heads, usually about one-half inch in diameter, are rose purple to pink or, rarely, white. The bracts beneath the flowers are *not* spiny. The fruits are oblong, flattened and somewhat curved, smooth and dark brown, and about one-eighth of an inch long.

This is one of the most troublesome thistles of cultivated fields, but it also grows in pastures and waste places, preferring heavy, moist soil.

This plant may be distinguished from other thistles by the almost spineless heads of unisexual flowers on green, wingless stems.

Compositae
(Composite family)
Tribe Cynareae

Cirsium plattense   Platte thistle (Plate 252)

This yellow-flowered thistle overwinters in the rosette stage. The early leaves may be entire or only wavy-margined, but later

128

leaves are successively more deeply cut. The leaves are white and velvety on the lower surface and grayish on the upper surface. The leaves extend as wings down the stem from their point of attachment. The lobes of the leaves terminate in a short yellow spine and weaker spines occur along the margin.

There are several flower heads on the Platte thistle; many of them terminate lateral branches which arise from the axils of the upper leaves of the main stem. The flowers are yellowish and each bract below the flowers is tipped with a short, yellow spine. The fruits are oval, straw-colored, and slightly curved.

These thistles may be found in central and western Nebraska growing in light sandy soil of the dry uplands, in overgrazed pastures, and in waste places. This is one of the first thistles to bloom. Flowering begins in late May and continues into June.

Compositae
(Composite family)
Tribe Cynareae

Cirsium vulgare   Bull thistle (Plate 253)

The stems of the bull thistle are winged with lobed, prickly leaf bases that run down the stem. The surface of the stem is more or less hairy. Branching in the upper third of the plant produces many stems which are leafy up to the heads.

The leaves are short (less than a foot long) and broad and very wavy or crinkled. Each lobe may have three or four points, each ending in a long yellow spine. Mature leaves arise alternately on the stem. The upper surface of the leaves may have short stiff hairs and frequently spines.

The heads may be solitary on the ends of branches, or two or three may occur together. The heads are filled with reddish purple flowers. The bracts are hairy and each terminates in a rigid yellow spine. The whole head seems constricted to about one-half of its diameter just below the corollas by the upper bracts of the inflorescence.

These plants inhabit pastures, roadsides, and waste places throughout Nebraska, but like the tall thistle it cannot stand cultivation.

129

Compositae
(Composite family)
Tribe Cichorieae

Cichorium intybus   Chicory (Plate 254)

Chicory is a cosmopolitan weed which is common in waste places and along roads in eastern and central Nebraska. The dried roots of this plant are often used as an additive in coffee.

The plants are perennial, the stems arising annually to over four feet in height from a deep taproot. The elongate and incised leaves of the plant base are similar to dandelion leaves. The upper leaves are much smaller, sessile, and sometimes nearly entire.

On any given day during the flowering season one can see one to four stalkless heads of clear blue ray flowers, produced in the axils of the small upper leaves. Under a hot sun, the heads close early in the day.

The end of each ray floret shows five teeth which indicate the five petals which were united to form the single strap-like corolla. There are no disk flowers in the head, and this condition holds for all members of this tribe, which includes the familiar dandelion.

Flowering is most common in July and August but may continue sporadically until October.

Compositae
(Composite family)
Tribe Cichorieae

Crepis runcinata   Hawk's-beard (Plate 255)

Some authors have chosen to call this species by the common name of dandelion hawk's-beard because of the similarity in the shape of the basal leaves to those of the dandelion. The specific name, *runcinata,* is from the word runcinate.

This species, a root perennial having one or more strong roots, is found in moist, open soil. Most of the leaves are basal and all are less than a foot long. They may bear hairs or lack them. The stem leaves are few in number and much smaller than the basal leaves. The stem is branched near the top with each branch terminating in a single flower head.

The flower heads bear many yellow flowers, each having strap-shaped corollas which are usually referred to as ligulate. This presents an appearance much like that of the head of a dandelion, but the heads are a little less than an inch in diameter. Supporting the flowers is a single row of nonoverlapping bracts which themselves are subtended by a series of much shorter bracts. The bracts are usually glandular-hairy, and in some plants they appear woolly. The fruits which are round in cross section bear a pappus composed of many soft white bristles.

These plants are usually under fifteen inches tall, but some may attain a height of two feet. This is the most common species of *Crepis* in the state and is sometimes used in cytology classes because of the small number of chromosomes in the nucleus of each cell.

Compositae
(Composite family)
Tribe Cichorieae

Lactuca pulchella   Blue lettuce (Plate 256)

Blue lettuce inhabits the prairie region of the United States and extends on westward. It is a perennial with deep rootstocks and stems up to three feet tall. This species is frequent throughout Nebraska in wet meadows, prairies, stream banks, and moist borrow pits of highways and railroads.

The stems lack hairs, or young stems may have hairs which are lost as the stems mature. The leaves are sessile and often entire, but the lower leaves are usually runcinate and deeply incised. The lanceolate leaves, which may be as much as seven inches long, tend to be covered with a whitish bloom of wax and have narrow tips.

The branches of the compound inflorescence are narrow. The heads, which are few in number, bear eighteen to fifty blue ligulate flowers. The heads are larger, becoming over an inch wide, and more showy than other species of *Lactuca.* The fruits produced by the flowers are about one-fourth of an inch long and taper into a beak of intermediate length compared to other species. The fruits

131

are thin and flat with several prominent striae or ridges on each surface. Flowering usually occurs from July to September.

Compositae
(Composite family)
Tribe Cichorieae

Lygodesmia juncea   Skeleton-weed (Plate 257)

The name rush pink has also been used as a common name for this plant. The cylindrical green, essentially leafless stems resemble rush stems to some people. The flowers more nearly resemble a garden pink than they do a composite head, and thus the alternate common name. The specific name, *juncea,* also refers to the generic name of the rushes, *Juncus.* To many of us who have seen this in the prairie and have been surprised by its apparent leaflessness and its ascending, branching, straggling habit of growth the name skeleton-weed seems particularly appropriate.

The stems are glabrous and often infected with an organism which produces globose galls. The upper stem has very much reduced scale-like leaves in the areas of stem branching. The lower leaves are an inch or two long, linear, and rigid.

Erect heads terminate the branches. Subtending the flowers in the form of a cylinder are four to eight bracts plus a few much reduced outer ones. There are usually five pale purple ligulate flowers per head. Each corolla or ligule has five teeth at the end; this is usually taken as an indication that five petals were united to make up the ligulate corolla. The flowers each turn outward so abruptly at the top of the involucre that a close look is necessary to see that there are five ligulate flowers instead of just five petals of a single flower. By close observation one can see the set of purple anthers making a tube through which the style grows. After emerging from the anthers the end of the style opens, exposing the two stigmatic surfaces. This is typical of nearly all fertile composite flowers. The fruits are elongate within the involucral bracts and are terminated by a set of soft tawny hairs making up the pappus.

These plants are common in prairies and along roadsides throughout the state, and they blossom from June to August.

132

Compositae
(Composite family)
Tribe Cichorieae

Microseris cuspidata   False dandelion (Plate 258)

The scientific name used for this plant in most manuals is *Agoseris cuspidata,* but Gleason and Cronquist (1963) use the name listed above. Many years ago Rydberg (1932) used the name *Nothocalais cuspidata.*

This is one of our early-blooming composites. It grows over most of the state in prairies as a perennial having a strong taproot. The unbranched, leafless flower stalks appear from April to June. The narrow, acuminate leaves arise from the crown of the taproot, and when young they are woolly on the entire, wavy margins.

The flower stalk at maturity may vary from two to twelve inches tall. The head bears many yellow ligulate flowers which are subtended by bracts of nearly equal length occurring in two or three series. The bracts are lance-shaped and have long slender tips. The fruits lack a beak and are shorter than the pappus which is composed of forty to fifty rigid, unequal capillary bristles.

Compositae
(Composite family)
Tribe Cichorieae

Taraxacum officinale   Dandelion (Plate 259)

Dandelion comes from the French *dent de lion,* which means lion's tooth. Dandelions are weeds not only in Nebraska but almost world-wide in temperate climates. They have been observed in blossom in almost every month of the year, but one of the largest flushes of blossoms occurs in April and May.

They are perennial from taproots, on the crowns of which appear a rosette of basal leaves which may even overwinter. The flower stalk does not bear leaves but varies considerably in height and bears a large head from one to two inches in diameter. The involucral bracts are in two series, the outer series shorter and the inner series reflexed. In fruit the bracts reflex and the mature olive brown achenes and pappus of white bristles at the end of a long beak form a conspicuous ball.

133

Closely mowed lawns and disturbed areas are common habitats for these plants.

Taraxacum laevigatum, the red-seeded dandelion, grows in eastern Nebraska. The fruits or achenes are red in this species, and the leaves are more slender than in the above species.

Compositae
(Composite family)
Tribe Cichorieae

Tragopogon dubius   Yellow goat's beard (Plate 260)

One of the beauties of the prairies which is missed by those who don't venture out until afternoon is the yellow goat's beard. It opens in bright sunlight and may close by noon. On cloudy days few if any of the flowering heads are open.

These are biennials which may grow as tall as two or two and one half feet when located in tall grass or in other taller vegetation. They bear fleshy, linear leaves that clasp the stem at the base and taper into a long narrow point resembling a grass leaf. They are alternate on the stem and continue upward until close to the terminal flowering head. Hairiness is common in young leaves.

The flower head is composed of many yellow ligulate flowers surrounded by ten to fourteen bracts which are longer than the ligulate flowers. The stems are swollen and hollow just below the head.

Roadsides and other open dry areas are the usual habitat of these plants. Flowering occurs from May to July, but usually the peak of flowering is in June. This scientific name is replacing *Tragopogon major* of earlier manuals.

Tragopogon pratensis is also a yellow-flowered goat's beard. The usual eight bracts of the flower head are equal to or shorter than the ligulate flowers, and the stem is not swollen and hollow beneath the head.

Plate 240   Townsendia
Townsendia grandiflora

Plate 241   Pussy-toes
Antennaria neglecta

Plate 242   Boneset
Eupatorium perfoliatum

Plate 243   False boneset
Kuhnia eupatorioides

Plate 244   Rough gay-feather
Liatris aspera

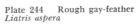

Plate 245   Dotted gay-feather
Liatris punctata

Plate 246   Gay-feather
Liatris pycnostachya

Plate 247   Blazing star
Liatris squarrosa var. *glabrata*

Plate 249   Musk thistle
Carduus nutans

Plate 248   Ironweed
Vernonia fasciculata

Plate 250   Tall thistle
Cirsium altissimum

Plate 251   Canadian thistle
Cirsium arvense

Plate 253   Bull thistle
Cirsium vulgare

Plate 252   Platte thistle
Cirsium plattense

Plate 254   Chicory
Cichorium intybus

Plate 255 Hawk's-beard
Crepis runcinata

Plate 256 Blue lettuce
Lactuca pulchella

Plate 257 Skeleton-weed
Lygodesmia juncea

Plate 258 False dandelion
Microseris cuspidata

Plate 259 Dandelion
Taraxacum officinale

Plate 260 Yellow goat's beard
Tragopogon dubius

References

Agricultural Research Service. 1970. Selected weeds of the United States. Agriculture Handbook No. 366. Washington: United States Department of Agriculture.

Anderson, Kling L., and Owensby, C. E. 1969. Common names of a selected list of plants. Bulletin 117. Manhattan, Kansas: Agriculture Experiment Station.

Bailey, L. H. 1949. Manual of cultivated plants. New York: Macmillan.

Barkley, T. M. 1968. A manual of the flowering plants of Kansas. Manhattan: Kansas State University Endowment Assoc.

Beal, E. O. 1956. Taxonomic revision of the genus *Nuphar* of North America and Europe. Jour. Elisha Mitchell Sci. Soc. 72:317-346.

Booth, W. E., and Wright, J. C. 1959. Flora of Montana. Bozeman: Montana State College.

Clements, F. E., and Clements, Edith S. 1914. Rocky Mountain flowers. Reprint. New York: Hafner Publishing Co., 1963.

Craighead, John J.; Craighead, Frank C., Jr.; and Davis, R. J. 1963. A field guide to Rocky Mountain wildflowers. Boston: Houghton Mifflin Co.

Dana, Mrs. W. S. 1893. How to know the wild flowers. Reprint. New York: Dover Publications, 1963.

Fassett, N. C. 1940. A manual of aquatic plants. New York: McGraw-Hill.

Fernald, M. L. 1950. Gray's manual of botany. 8th ed. New York: American Book Co.

Gates, Frank C. 1932. Wild flowers in Kansas. Topeka: Kansas State Board of Agriculture.

Gleason, Henry A. 1968. The new Britton and Brown illustrated flora. 3 vols. New York: Hafner Publishing Co.

Gleason, H. A., and Cronquist, Arthur. 1963. Manual of vascular plants. Princeton: Van Nostrand.

Harrington, H. D. 1964. Manual of the plants of Colorado. 2d ed. Denver: Sage Books.

Johnson, J. R., and Nichols, J. T. 1970. Plants of South Dakota grasslands. Brookings, S. D.: Agricultural Experiment Station.

Justice, W. S., and Bell, C. Ritchie. 1968. Wild flowers of North Carolina. Chapel Hill: University of North Carolina Press.

Kingsbury, J. M. 1964. Poisonous plants of the United States and Canada. Englewood Cliffs: Prentice-Hall.

Lemmon, R. S., and Sherman, C. L. 1958. Flowers of the world. Garden City: Hanover House.

Marks, R. H. 1962. Nebraska weeds. Dept. of Agr. and Insp. Bulletin No. 101-R. Lincoln, Nebraska: Weed and Seed Division.

McCarty, M. K.; Scifrcs, C. J.; and Robison, L. R. 1967. A descriptive guide for major Nebraska thistles. SB 493. Lincoln, Nebraska: Agricultural Experiment Station.

Nebraska Centennial Commission. 1967. Common and scientific names of a selected list of Nebraska plants. Lincoln: University of Nebraska Department of Information.

Nelson, Ruth A. 1969. Handbook of Rocky Mountain plants. Tucson: D. S. King Publisher.

Novak, F. A. 1966. The pictorial encyclopedia of plants and flowers. New York: Crown Publishers.

Omaha Botany Club. 1959. Plants of Fontenelle Forest. Omaha: Fontenelle Forest Assoc. and Omaha Botany Club.

Petersen, N. F. 1923. Flora of Nebraska. 3d ed. Lincoln, Nebraska: The Author.

Peterson, R. T., and McKenny, M. 1968. A field guide to wild flowers. Boston: Houghton Mifflin Co.

Pool, R. J. 1948. Marching with the grasses. Lincoln: University of Nebraska Press.

Pool, R. J., and Maxwell, Earl G. 1952. Wild flowers in Nebraska. EC 5-168. Lincoln: University of Nebraska College of Agriculture Extension Service.

Rickett, H. W. 1966. Wild flowers of the United States. New York: McGraw-Hill.

Rydberg, P. A. 1932. Flora of the prairies and plains of central North America. New York: New York Botanical Garden.

Stevens, W. C. 1948. Kansas wild flowers. Lawrence: University of Kansas Press.

Steyermark, Julian A. 1963. Flora of Missouri. Ames: Iowa State University Press.

Van Bruggen, Theodore. 1971. Wildflowers. Interior, S. D.: Badlands Natural History Assoc.

Waterfall, U. T. 1958. A taxonomic study of the genus *Physalia* in North America north of Mexico. *Rhodora* 60:107-114, 128-142, 152-173.

Weaver, J. E. 1954. North American prairie. Lincoln, Nebraska: Johnsen Publishing Co.

Weaver, J. E. 1965. Native vegetation of Nebraska. Lincoln: University of Nebraska Press.

Welsh, S. L., and Ratcliffe, Bill. 1971. Flowers of the canyon country. Provo, Utah: Brigham Young University Press.

Winter J. M. 1936. An analysis of the flowering plants of Nebraska. Lincoln: Nebraska Conservation and Survey Division.

Aske Whitebird (Rosebud Sicangu), Office of SD Tribal Govt. Relations, at a SDIBA mtg. in Pierre, SD.

Coya Night Pipe (Rosebud Sicangu), Human Resource Generalist, Rapid City Area Schools.

Dist. 27 **Senator Jim Bradford** on hiatus in Rapid City, SD.

Deborah Gangloff is president and CEO of Blue Rabbit, a consulting firm specializing in services to non-profit organizations including grant writing/development, publications, planning and governance. She worked in the museum field for thirty years, specializing in collections management in New York, Chicago and South Dakota before starting her own business. She lives in Deadwood, SD.

Sandra Abena Songbird (Naylor) is an Abenaki Indian/French Canadian/Irish poet, writer and singer; a member of the Missisquoi Abenaki of Swanton, Vermont. She is the author of "Bitterroot" (Freedom Voices Press) a collection of poetry, and CD, "They're Calling Us Home." Former Senior Staff Writer for the Dakota Lakota Journal (2005-2008). Abena currently is a freelance writer/photographer submitting to Native Legacy Magazine.

Lila DeMarrias Mehlhaff is from Little Eagle on Standing Rock Reservation and, has been married to Stew Mehlhaff for 23 years. They have 4 children, 2 cats and a Boston Terrier together. They love living in the Black Hills of South Dakota and enjoy going to their kids' sporting events, eating Elks Theater popcorn and watching the movie Castaway over and over and over again.

By Deborah Gangloff

Forty-one years ago when Father Ted Zuern, S.J. and Bob Savage of Omaha had the idea to hold a small art show at the Red Cloud School, one wonders if they had any idea what they were letting loose in the world. From a modest beginning in 1968, the annual Red Cloud Indian Art Show has grown to be one of the most important incubators in the nation for fledgling and emerging artists, as well as a reliable support base for some of the finest artists in the Northern Plains region.

Through it, The Heritage Center at Red Cloud School has accumulated an art collection that is the envy of many museums. It also defined the career and was the life's work of Brother Simon, S.J. one of one of Red Cloud's most beloved and unforgettable characters.

Brother Simon, a bookkeeper/office manager at the school whose responsibilities included the management of the school's herd of Charolais ("Simon used to say he wcs secretary to thirty tall blondes," said John Day, his long time friend and colleague and former dean of the School of Art at University of South Dakota), assisted Father Zuern for just a few years before he himself was put in charge of the show.

Brother Simon did not know much about art, but he had good instincts, learned quickly, and worked hard in his role as Director. He made it a priority to learn the field, visiting exhibits and shows, talking with collectors and establishing relationships with artists. Like Father Zuern, he saw the show as a way to support Red Cloud School's mission to promote Native culture. He also saw it as an opportunity to establish an important art collection through purchases from the show.

"He developed an excellent eye. He watched the development of artists' careers, buying at one phase and then waiting, buying again when they changed direction" said Mr.

artists, and the Red Cloud Show was – and is – the perfect place for the savvy collector to find an exciting new artist or a new direction from an old favorite who decided to experiment with something at Red Cloud.

The show today is organized and mounted by The Heritage Center's Curator Mary Bordeaux, Director Peter Strong, and assisted by Miriam Norris. It remains, as it was under Brother Simon's direction, a supportive environment where the country's most exciting new artists may get off to a good start, as well as a place to find work by some of the biggest names in Northern Plains art.

Don Montileaux, Don Ruleaux, Roger Broer, Andrea Two Bulls, and Martin Red Bear

Allison, and Denton Lafferty are finding the Red Cloud show to be a valuable venue to gain exposure and experience.

Famed artist Don Montileaux began his career at Red Cloud, exhibiting his work for the first time there in 1969. Montileaux, who still exhibits at Red Cloud even though he has an established, successful career, recognizes Red Cloud as the ideal place for "young artists to see how seasoned artists present their work." He also cites it as a "good seller".

Indeed, both Mary Bordeaux and John Day remarked on the predictably high percentage of usually 35% - 50% sales from the show.

The Red Cloud Indian Art Show is an on-going, year-round

Market and others, to meet artists, establish relationships, become familiar with who is showing, and to spread the word about the Red Cloud annual show.

Bordeaux said they are always looking to see "where the buzz is - who people are talking about."

The Red Cloud Indian Art Show is one of the last intertribal art shows in the nation, and its location is important as one of the few Native art shows located within a reservation. The call for submissions usually goes out at the end of January, and the deadline to get artwork to Red Cloud is the third week in May. All tribally-enrolled artists who are at least 18 years of age who submit work are accepted,

Keys to Plant Genera

A *Monocotyledons; leaves with striate venation; stems with scattered vascular bundles; flowers usually have parts in threes.*

B Flowers without petals; perianth represented by scales or bristles, or lacking.

   C Plants aquatic or found in swampy areas.

      D Flowers in spherical heads on branches arising from leaf axils; fruits apparent **Sparganium** (bur-reed)

      D Flowers in terminal cylindrical spikes, the upper staminate drying and disappearing; fruit hidden among bristles .... **Typha** (cattail)

   C Plants terrestrial found in moist woods; leaves basal with 3 leaflets; inflorescence a fleshy spike surrounded by a large bract which arches over it .... **Arisaema** (jack-in-the-pulpit)

B Flowers with petals; sepals sometimes also petal-like.

   E Flowers regular.

      F Pistils several to many, separate.

         G Pistils in a ring on a flat receptacle .... **Alisma** (water-plantain)

         G Pistils covering a convex receptacle .... **Sagittaria** (arrow-head)

      F Pistils one, but may have one or more cells.

         H Stamens 6.

            I Flowers in groups arising from leaf-like bracts .... **Tradescantia** (spiderwort)

            I Flowers not borne as above.

               J Ovary superior .... **Liliaceae** (see key below)

               J Ovary inferior .... **Hypoxis** (yellow star-grass)

         H Stamens 3.

            K Plants growing in mud in marshy areas; flower stalk not flattened; ovary superior .... **Heteranthera** (mud-plantain)

K Plants growing in prairie sod; flower stalk flattened;
    ovary inferior .... **Sisyrinchium (blue-eyed grass)**
E Flowers irregular.
  L Petals 3, 2 large blue and 1 small white; ovary superior .... **Commelina (day-flower)**
  L Petals 3, modified but not as above; ovary inferior.
    M Plants without green leaves .... **Corallorhiza (coral-root)**
    M Plants with green leaves.
      N Lower petal a large sac .... **Cypripedium (lady-slipper)**
      N Lower petal different from others but not a sac.
        O Plants with two basal leaves; flowers showy, 3 to 6 on a stalk ...... **Orchis (showy orchis)**
        O Plants with more than 2 leaves; flowers small white, many,
          in 3 vertical or spiral rows on a stalk .... **Spiranthes (ladies' tresses)**

Liliaceae, key to genera.

a Sepals and petals unlike in color and/or shape.
  b Leaves several, alternate, narrow .... **Calochortus (sego lily)**
  b Leaves 3, whorled, ovate .... **Trillium (trillium)**
a Sepals and petals nearly alike.
  c Leaves basal, not on an erect stem.
    d Leaves long, sharp pointed, rigid, evergreen .... **Yucca (yucca)**
    d Leaves not as above, not rigid nor evergreen.
      e Leaves broad, usually 2, often mottled .... **Erythronium (fawn lily)**
      e Leaves narrow, usually several to many, green.
        f Flowers several on a stalk in an umbel .... **Allium (onion)**
        f Flowers solitary from the ground, not on a stalk .... **Leucocrinum (sand-lily)**
  c Leaves not all basal, some or all borne on a stem.
    g Leaves whorled .... **Lilium (lily)**

g Leaves not whorled.

  h Flowers in several small clusters drooping

    from the axils of broad leaves ....**Polygonatum (Solomon's seal)**

  h Flowers in terminal racemes or panicles.

    i Leaves long, narrow, grass-like near the base, reduced above. ...... **Zigadenus (death camas)**

    i Leaves broad, alternate not clustered near the base,

      not reduced conspicuously above .... **Smilacina (false Solomon's seal)**

A *Dicotyledons; leaves with net venation; stems with vascular bundles in a ring; flowers usually have parts in fours or fives.*

P Flowers without true petals; petal-like organs (colored sepals) usually present. ....**Dicot Key 1**

P Flowers with true petals; sepals or pappus usually present also.

  Q Flowers with several separate petals or petal-like organs.

    R Flowers regular.

      S Pistils several and separate in each flower. ....**Dicot Key 2**

      S Pistils one in each flower.

        T Ovary superior.

          U Stamens numerous, at least more than twice the number of petals. ....**Dicot Key 3**

          U Stamens not more than twice the number of petals. ....**Dicot Key 4**

        T Ovary inferior. ....**Dicot Key 5**

    R Flowers irregular; in many legumes the lower two petals are fused to form a keel. ......**Dicot Key 6**

  Q Flowers with petals fused at least at the base, often forming a tube.

    V Ovary superior.

      W Flowers regular. ....**Dicot Key 7**

      W Flowers irregular. ....**Dicot Key 8**

    V Ovary inferior.

X Flowers not in heads having an involucre of bracts. ....**Dicot Key 9**
X Flowers in heads which are subtended by an involucre of bracts. ....**Dicot Key 10**

Dicot Key 1. Herbaceous plants having flowers lacking true petals. In certain flowers (*Mirabilis*) the sepals may appear to be petals and the involucre may appear to be sepals. In others (*Euphorbia*) the appendages of a glandular involucre may appear to be petals. In most flowers of the dicotyledons if there is but a single whorl of perianth parts, it is usually considered to consist of sepals, the petals being absent.

A Ovary inferior...........................**Comandra (bastard toadflax)**

A Ovary superior.

  B Pistils more than one per flower.

    C Calyx of separate, often petal-like sepals; pistils many.

      D Climbing vine; leaves opposite.............**Clematis (clematis)**

      D Herbaceous plants; leaves mostly basal........**Anemone (anemone)**

    C Calyx cup-like; pistils five, united below..........**Penthorum (penthorum)**

  B Pistils one in a flower, simple or of several carpels.

    E Leaves with conspicuous teeth or lobes.

      F Plants producing latex (milky juice); flowers in a cup-like glandular involucre. **Euphorbia (spurge)**

      F Plants not producing latex; flowers not in a glandular cup-like involucre.

        G Leaves persistent, broadly ovate with serrate margins and without stinging hairs..............**Laportea (wood nettle)**

        G Leaves soon deciduous, lanceolate, sinuate toothed and without stinging hairs.............**Cycloloma (winged pigweed)**

    E Leaves with entire or undulate margins, teeth, if present, very fine.

      H Fruit a berry with red juice..............**Phytolacca (pokeberry)**

      H Fruit dry.

142

I Fruit a three-seeded capsule; in most species the flowers much reduced, the involucrate cluster of flowers easily mistaken for a flower; latex present.. **Euphorbia** (spurge)

I Fruit an acheme or utricle.

  J Fruit a lens-shaped or triangular achene; stipules usually united and sheathing, that is having an ocrea. If no ocreae are present then the flowers are subtended by an involucre.

    K Ocreae absent; flowers with an involucre.................... **Eriogonum** (eriogonum)

    K Ocreae present; flowers without an involucre.

      L Outer sepals spreading in fruit, the inner winged and 3/4 inch or more wide; stigmas tufted................... **Rumex** (dock)

      L Outer sepals erect in fruit, the inner never winged; stigmas capitate.................................... **Polygonum** (smartweed)

  J Fruit a utricle; stipules, if present, not forming ocreae.

    M Calyx corolloid, the limb deciduous, the base persistent around the fruit.

      N Flower cluster subtended by an involucre of united bracts; fruit ribbed.............................. **Mirabilis** (four-o'clock)

      N Flower cluster subtended by an involucre of distinct bracts; fruit winged.......................... **Abronia** (sand-verbena)

    M Calyx not corolloid; green or scarious; plants with woolly terminal spikes.................................. **Froelichia** (froelichia)

Dicot Key 2. Plants bearing regular polypetalous flowers, each having several pistils.

A Flowers having spurs... **Aquilegia** (columbine)

A Flowers without spurs.

  B Plants living in water or in marshy land...... **Ranunculus** (buttercup, crowfoot)

  B Plants not living in water or in marshes.

143

    C Fruit a strawberry; all leaves bearing only three leaflets..Fragaria (strawberry)
    C Fruit not a strawberry; most leaves having more than three leaflets.
        D Stems with prickles ....Rosa (rose)
        D Stems without prickles ....Potentilla (cinquefoil)

Dicot Key 3. Herbaceous plants with regular, polypetalous flowers having one pistil per flower. Ovary superior. Stamens numerous, at least more than twice the number of petals.

A Stamens united by their filaments into a central column surrounding the pistil, united with the petal bases.
    B Stamen-column bearing anthers along the sides; fruit a capsule........................Hibiscus (hibiscus)
    B Stamen-column bearing anthers at the summit; fruit of several carpels in a ring around a central axis.
        C Stigma terminal, capitate..Sphaeralcea (globe mallow)
        C Stigma linear on the side of the thread-like style branches.
            D Petals notched at the apex; carpels beakless...Malva (mallow)
            D Petals not notched; carpels beaked...Callirhoe (poppy-mallow)
A Stamens not united.
    E Leaves opposite...Hypericum (St. John's-wort)
    E Leaves not opposite.
        F Plants aquatic; leaves floating or emersed; petals and stamens numerous.
            G Flowers yellow; leaves emersed..Nuphar (yellow pond-lily)
            G Flowers white; leaves floating...Nympheae (water-lily)*
        F Plants not aquatic; leaves lobed, juice red or yellow.
            H Leaves all basal, without spines; juice red...................................Sanguinaria (blood-root)
            H Leaves arising from the aerial stem, with spines; juice yellow................Argemone (prickly poppy)

*Nympheae has an inferior ovary, but is keyed here for convenience and contrast.

144

Dicot Key 4. Herbaceous plants with regular, polypetalous flowers having one pistil in each. Ovary superior. Stamens not more than twice the number of petals.

A Calyx or receptacle cup-like.
  B Flowers inconspicuous in small axillary clusters..**Ammannia (tooth-cup)**
  B Flowers conspicuous, solitary in upper axils, dimorphic.....................**Lythrum (winged loosestrife)**
A Calyx or receptacle not cup-like.
  C Plants succulent...**Claytonia (spring beauty)**
  C Plants not succulent.
    D Flowers in spherical heads.
      E Flowers white; plants erect herbs...**Desmanthus (prairie mimosa)**
      E Flowers purple; plants scrambling vines..**Schrankia (sensitive briar)***
    D Flowers not in spherical heads.
      F Leaves opposite or whorled and entire..**Carophyllaceae (see key below)**
      F Leaves alternate or basal.
        G Stamens 6.
          H Petals and sepals 6..**Podophyllum (May apple)**
          H Petals 4; sepals 4–8.
            I Stamens in two groups, 2 shorter than the others.............**Cruciferae (see key below)**
            I Stamens all equal, exserted..**Cleome (cleome)**
        G Stamens 5–10; petals and sepals 5.
          J Fruit with stout spines; plants prostrate with pinnately
             compound leaves...**Tribulus (puncture vine)**
          J Fruit or ovary not spiny; plants not prostrate, but some
             may have pinnately compound leaves.
            K Leaves simple or pinnately compound.
              L Leaves simple...**Linum (flax)**
              L Leaves pinnately compound..**Cassia (partridge-pea)**

* Although *Schrankia* is sympetalous not polypetalous it is keyed here for convenience and contrast.

145

K Leaves palmately divided or compound.
  M Leaves 3-foliate, heart-shaped...Oxalis (wood-sorrel)
  M Leaves palmately lobed or divided............................Geranium (geranium)

Caryophyllaceae, key to genera.

  a Calyx of distinct sepals.
    b Petals deeply 2-parted...Stellaria (chickweed)
    b Petals entire or emarginate.....................................Arenaria (sandwort)
  a Calyx of united sepals
    c Styles 3 (or 4)..Silene (catchfly)
    c Styles 2 (or 3).
      d Flowers many in bloom at one time, conspicuous.........Saponaria (bouncing-Bet)
      d Flowers few in blossom at one time, inconspicuous.......Dianthus (pink)

Cruciferae, key to genera.

  a Pods flattened perpendicular to the narrow partition.......Thlaspi (penny-cress)
  a Pods not flattened perpendicular to the partition.
    b Pods flattened parallel to the broad partition................Dentaria (toothwort)
    b Pods round or 4-sided in cross-section.
      c Pods short (usually less than twice as long as wide).
        d Flowers white...Berteroa (false alyssum)
        d Flowers yellow...Lesquerella (bladder pod)
      c Pods long (several times as long as wide).
        e Pod with a distinctly flattened beak....................Brassica (mustard)
        e Pod not beaked.

  f Pods 4-sided in cross-section. .... Erysimum (erysimum)

  f Pods round in cross-section.

   g Seeds in 2 rows in each cell of the pod; aquatic. .... Rorippa (water cress)

   g Seeds in 1 row in each cell of the pod; terrestrial.

    h Flowers yellow. .... Sisymbrium (sisymbrium)

    h Flowers purple. .... Hesperis (dame's rocket)

Dicot Key 5. Herbaceous plants with regular, polypetalous flowers having one pistil per flower. Ovary inferior.

A Plants with floating leaves or spiny stems.

 B Leaves floating, round; flowers large white. .... Nymphaea (water lily)

 B Leaves mostly absent; stem spiny; flowers colored.

  C Stem globose, covered with tubercles which have clusters of spines

   at the apex. .... Coryphantha (ball cactus)

  C Stem jointed, the joints flattened or cylindric, with spines and usually

   small barbed bristles. .... Opuntia (prickly pear)

A Plants with neither floating leaves nor succulent stems.

 D Flowers in umbels; petals 5. .... Umbelliferae (see key below)

 D Flowers not in umbels; petals 4 to many.

  E Stamens many; petals 5 or 10. .... Mentzelia (mentzelia)

  E Stamens as many or twice as many as the 4 petals.

   F Fruit indehiscent; flowers red. .... Gaura (gaura)

   F Fruit a many-seeded capsule; flowers white, pink or yellow.

    G Seeds with a tuft of hairs. .... Epilobium (willow-herb)

    G Seeds without a tuft of hairs. .... Oenothera (evening primrose)

147

Umbelliferae, key to genera.

a Leaves palmately divided into 3 ovate leaflets..Heracleum (cow-parsnip)

a Leaves pinnately once or more lobed or divided.

  b Flowers yellow.

    c Plants stemless, scapes and leaves arising from the fleshy root..................Lomatium (parsley)

    c Plants with leafy stems; leaves pinnately divided into ovate, lobed,
      and toothed segments...Pastinaca (parsnip)

  b Flowers white, greenish white or occasionally pink or purple.

    d Plants without stems...Lomatium (parsley)

    d Plants with leafy stems.

      e Fruits covered with spines; central flower of umbel often purple.............Daucus (carrot)

      e Fruits not covered with spines.

        f Leaves simple pinnate, leaflets linear to lanceolate or finer.......Sium (water parsnip)

        f Leaves 2–4 pinnate, leaflets lanceolate to oblong.

          g Plants up to six feet tall; principal leaves 2–3 pinnately
            compound; wet places...Cicuta (water-hemlock)

          g Plants with purple splotched stems up to nine feet tall; principal
            leaves 3–4 pinnately compound; waste places...................Conium (poison-hemlock)

**Dicot Key 6. Herbaceous plants with irregular, polypetalous flowers; in many legumes the lower petals fuse to form
a keel.**

A Flowers with evident spurs.

  B Spurs produced by both sepals and petals...Delphinium (larkspur)

  B Spurs produced by sepals or petals but not both.

    C Spur one, on the upper of three sepals; corolla 2-lipped orange or yellow. Impatiens (touch-me-not)

    C Spur from petals.

D Spur from the lower of five petals; stamens 5. .... **Viola (violet)**

D Spur from outer petals; stamens 6.

    E Spurs one, from one outer petal; flowers yellow. .... **Corydalis (corydalis)**

    E Spurs two, from both outer petals which are fused at the base

        appearing heart-shaped; flowers white or pinkish. .... **Dicentra (Dutchman's breeches)**

A Flowers without spurs.

  F Calyx of united sepals; flower parts in fives.

    G Lower petal enlarged and sac-like; stamens five. .... **Viola (violet)**

    G Lower two petals united along lower edge to form a boat-like keel

        which encloses the pistil and usually ten stamens. .... **Leguminosae** (see key below)

  F Calyx of five separate, unequal sepals, two of which are large and petal-like;

    petals 3, somewhat fused. .... **Polygala (milkwort)**

Leguminosae, key to genera.

  a Leaflets three. (Foliaceous stipules on nearly sessile leaves may give the appearance of five leaflets. Contrary-

    wise, the five leaflets of *Lotus* look like three leaflets and two basal stipules.)

    b Stamens not united by their filaments.

      c Pods up to 2 inches long, inflated; flowers white or cream colored. .... **Baptisia (wild indigo)**

      c Pods up to 4 inches long, flat; flowers yellow. .... **Thermopsis (golden-pea)**

    b Stamens united by their filaments into one or two groups.

      d Pod a loment. .... **Desmodium (tick trefoil)**

      d Pod not a loment.

        e Foliage glandular dotted.

          f Leaves palmately compound; pubescent. .... **Psoralea (scurf-pea)**

          f Leaves pinnately compound; glabrous. .... **Petalostemon (prairie-clover)**

        e Foliage not glandular dotted.

          g Pods curved or coiled. .... **Medicago (alfalfa)**

          g Pods not curved or coiled.

h Flowers in heads. .... **Trifolium** (clover)

h Flowers not in heads.

  i Flowers in elongate racemes. .... Melilotus (sweet-clover)

  i Flowers in umbels on long stalks. .... **Lotus (bird's foot trefoil)***

a Leaflets five or more.

  j Leaves palmately compound.

    k Foliage glandular dotted. .... Psoralea (scurf-pea)

    k Foliage not glandular dotted. .... Lupinus (lupine)

  j Leaves pinnately compound.

    l Flowers in spherical heads.

      m Flowers white; erect herbs without spines. .... **Desmanthus** (prairie mimosa)

      m Flowers purple; scrambling vines with recurved spines; leaves usually sensitive to touch. .... **Schrankia** (sensitive briar)

    l Flowers not in spherical heads.

      n Herbs; leaves odd pinnately compound.

        o Flowers in terminal clusters.

          p Flowers in umbels on long stalks. .... **Lotus (bird's foot trefoil)**

          p Flowers in spikes or racemes.

            q Petal one (the standard); plants woody at the base. .... Amorpha (lead plant)

            q Petals five; bases not woody; stamens five. .... Petalostemum (prairie-clover)

        o Flowers on axillary peduncles.

          r Pods prickly; leaflets glandular dotted. .... Glycyrrhiza (wild licorice)

          r Pods smooth; leaflets not glandular dotted.

            s Flowers not pea-like; petals yellow, slightly unequal in size; stamens 10, 4 yellow, 6 purple. .... **Cassia** (partridge-pea)

            s Flowers pea-like; in axillary racemes. .... **Astragalus** (milk-vetch)

* Keyed here for convenience (see note above under a).

n Vines; leaves even pinnately compound, rachis ending in a tendril or bristle.
    t Style slender, hairy at the apex. .... **Vicia (vetch)**
    t Style flattened toward the apex, hairy on the inner side. .... **Lathyrus (vetchling)**

Dicot Key 7. Herbaceous plants with regular, sympetalous flowers having superior ovaries.

A Stamens free from the corolla; non-green saprophytes. .... **Monotropa (Indian-pipe)**
A Stamens borne on the corolla, or if free, the plants green.
  B Stamens opposite the corolla lobes; leaves opposite or whorled, basal or cauline.
    C Corolla lobes reflexed, purple; leaves basal. .... **Dodecatheon (shooting star)**
    C Corolla lobes erect or spreading, yellow; leaves mostly from the stem. .... **Lysimachia (loosestrife)**
  B Stamens alternate the corolla lobes.
    D Ovaries two or united by their stigmas; juice milky.
      E Styles united; stamens distinct.
        F Flowers solitary; seeds without a tuft of hairs. .... **Vinca (periwinkle)**
        F Flowers in groups; seeds with a tuft of hairs. .... **Apocynum (dogbane)**
      E Styles distinct; stamens often fused to the united peltate stigmas. .... **Ascelepias (milkweed)**
    D Ovary one.
      G Ovary deeply four-lobed, the style arising from between the lobes. .. **Boraginaceae (see key below)**
      G Ovary not deeply four-lobed.
        H Ovary one-celled; fruit a capsule.
          I Leaves entire and opposite; corolla 4–6 lobed.
            J Corolla cup-shaped, purple; flowers large. .... **Eustoma (prairie-gentian)**
            J Corolla with plaits in the sinuses, blue; flowers hardly
              opening or vase-shaped. .... **Gentiana (gentian)**
          I Leaves 3-foliate, basal, segments not toothed; corolla
            white, bearded on the upper surface; marsh plants. .... **Menyanthes (buckbean)**

151

H Ovary two-celled or more.

 K Stamens 2 or 4; leaves opposite.

  L Stamens 2; ovary 2-celled; corolla lobes four..................**Veronica (speedwell)**

  L Stamens 4, in two groups; ovary 4-celled; corolla lobes 5............**Verbena (vervain)**

 K Stamens 5.

  M Vines or weak shrubs with large flowers and simple leaves.

   N Vines; stigmas 2 (or 3) threadlike; leaves arrow shaped.. **Convolvulus (bindweed)**

   N Large shrub; stigma globose; leaves linear.............**Ipomoea (morning glory)**

  M Herbs; leaves various.

   O Style one, unbranched; leaves alternate.

    P Plants large, with flowers in a dense terminal spike. **Verbascum (mullein)**

    P Plants small often prickly with flowers axillary
       or borne in cymes.............................**Solanaceae** (see key below)

   O Style one, stigmas 2 or 3.

    Q Stigmas 2; herbs with lobed or divided leaves. **Hydrophyllum (water leaf)**

    Q Stigmas 3.

     R Leaves simple, entire, mostly opposite leaves...........**Phlox (phlox)**

     R Leaves dissected into narrow segments, some alternate;
         flowers white, tubes long.........................**Gilia (gilia)**

Boraginaceae, key to genera.

 a Corolla irregular, the lobes unequal and the tube bent; stem spiny; flowers blue;
    stamens exserted. ....**Echium (blue-weed)**

 a Corolla regular.

   b Plants annual; leaves alternate; densely hairy; corolla white...........**Cryptantha (cryptantha)**

   b Plants perennial.

     c Corolla tubular, its lobes acute, erect; style long exserted; leaves
        lanceolate or wider.....................................**Onosmodium (false gromwell)**

152

   c Corolla funnel-form, its lobes rounded, spreading; style not conspicuous;
     leaves linear; roots red..**Lithospermum (puccoon)**

Solanaceae, key to genera.

   a Plants large, coarse; leaves large, coarsely toothed; flowers large, solitary;
     fruit a spiny capsule opening from the top...........................**Datura (Jimson-weed)**
   a Plants smaller; flowers in small clusters; fruit a berry.
     b Fruiting calyx becoming inflated and bladdery; flowers cup-shaped,
       yellowish with a dark center...**Physalis (ground-cherry)**
     b Fruiting calyx not enlarging in fruit; flowers yellow; plants often spiny........**Solanum (buffalo-bur)**

Dicot Key 8. Herbaceous plants with irregular, sympetalous flowers having superior ovaries.

A Corolla not truly sympetalous, with some petals joined, not forming a tubular flower. See Dicot Key 6 for
   such flowers as *Impatiens, Polygala, Corydalis, Dicentra,* and the family Leguminosae.
A Corolla truly sympetalous and tubular at least at the bottom. The corolla either 2-lipped
   or some corolla lobes larger than others. Certain species may appear regular.
   B Stamens with anthers 5..**Verbascum (mullein)**
   B Stamens with anthers 2–4.
     C Ovules solitary in the 2–4-celled ovary; leaves opposite or whorled.
       D Stamens with anthers 4, in pairs; flowers small in spikes.
         E Corolla 5-lobed, nearly regular; nutlets 4.....................................**Verbena (vervain)**
         E Corolla 4-lobed, 2-lipped; nutlets 2...**Phyla (fog-fruit)**
       D Stamens with anthers 2; flowers 2-lipped; stems usually square
         in cross-section...**Labiatae** (see key below)
     C Ovules 2-many in each cell of the ovary; leaves opposite or alternate.

153

F Plants parasitic, lacking chlorophyll; leaves scale-like; ovary 1-celled; stamens 4...**Orobanche (broom-rape)**

F Plants with green leaves; stamens 2 or 4.

  G Fruit a 1-celled capsule with 2 large incurved long beaks at the apex..**Proboscidea (unicorn-plant)**

  G Fruit a 2-celled capsule, without long beaks...............**Scrophulariaceae (see key below)**

Labiatae, key to genera.

a Anther-bearing stamens 4.

  b Flowers in terminal spikes.

    c Corolla split nearly to the base on the upper side.......................**Teucrium (wood sage)**

    c Corolla not split nearly to the base on the upper side.

      d Calyx nearly equally 5-toothed, not 2-lipped, plants medium to tall..........**Nepeta (catnip)**

      d Calyx not equally 5-toothed, 2-lipped, plants short...................**Prunella (heal-all)**

  b Flowers in axillary clusters, not in spikes.

    e Leaves much longer than broad..................................**Leonurus (motherwort)**

    e Leaves about as broad as long...................................**Glecoma (ground ivy)**

a Anther-bearing stamens 2.

  f Flowers in terminal spikes of few-flowered clusters.............................**Salvia (sage)**

  f Flowers in dense, bracted, head-like terminal or axillary clusters........**Monarda (wild bergamot)**

Scrophulariaceae, key to genera.

a Anther-bearing stamens 5; corolla nearly regular; plants large, woolly.........**Verbascum (mullein)**

a Anther-bearing stamens 2 or 4, the 5th, if present, sterile.

  b Corolla spurred at the base.....................................**Linaria (butter and eggs)**

  b Corolla not spurred at the base.

    c Anther-bearing stamens 4; capsule never flattened.

154

     d Stamens 5, but 1 sterile and nearly as long as the others.............**Pentstemon** (penstemon)

     d Stamens 4, all anther bearing.

        e Corolla nearly regular (hardly 2-lipped); stamens in 2 groups;

          leaves alternate, linear..............**Agalinis** (gerardia)

        e Corolla distinctly 2-lipped; leaves entire, opposite, roundish.......**Bacopa** (water-hyssop)

   c Anther-bearing stamens 2; capsules somewhat flattened; corolla

     almost regular...............**Veronica** (speedwell)

Dicot Key 9. Herbaceous plants with sympetalous flowers having inferior ovaries. Flowers not in involucrate heads.

A Flowers regular.

  B Stamens as many as the corolla lobes and alternate with them, separate.

    C Leaves mostly opposite or whorled.

      D Leaves always with stipules.

        E Plants perennial from a woody root; leaves opposite...............**Houstonia** (bluets)

        E Plants annual; leaves whorled..............**Galium** (bedstraw)

      D Leaves usually without stipules; a coarse weed with flowers

        in the axils of perfoliate leaves..............**Triosteum** (horse-gentian)

    C Leaves alternate, without stipules.

      F Corolla bell-shaped in solitary flowers or rotate in spicate flowers,

        present in all flowers..............**Campanula** (bell-flower)

      F Corolla rotate, wanting in most of the sessile, axillary flowers.....**Triodanis** (Venus' looking-glass)

  B Stamens united by their anthers; vines with unisexual flowers..............**Echinocystis** (wild cucumber)

A Flowers irregular, 2-lipped; leaves alternate..............**Lobelia** (lobelia)

Dicot Key 10. Herbaceous plants with sympetalous flowers having inferior ovaries. Flowers in involucrate heads.

Compositae, key to tribes.

A Flowers all ligulate, perfect; plants with milky juice................................**Cichorieae**

A Flowers all tubular, or heads with both tubular and ligulate flowers; ray flowers pistillate or sterile.

  B Receptacle densely bristly; true ray flowers never present, marginal flowers sometimes enlarged..**Cynareae**

  B Receptacle naked or chaffy, or if bristly, ray flowers present.

    C Involucral bracts green, or only the tip or margins scarious.

      D Receptacle naked or rarely bristly; pappus often capillary.

        E Involucral bracts imbricated in 2-many series.

          F Ray flowers wanting; flowers all perfect, never bright yellow.

            G Style branches filiform; leaves alternate; pappus never plumose nor heads spicate.................................**Vernonieae**

            G Style branches thick, obtuse; leaves opposite, or if alternate, pappus plumose or heads spicate............**Eupatorieae**

          F Ray flowers usually present, if wanting the flowers bright yellow or the marginal pistillate...........................**Astereae**

        E Involucral bracts little if at all imbricated.

          H Pappus of chaffy scales or absent...........................**Helenieae**

          H Pappus of capillary bristles..................................**Senecioneae**

      D Receptacle chaffy, each flower in the axil of a bract; pappus never of capillary bristles...**Heliantheae**

    C Involucral bracts scarious.

      I Pappus usually present, capillary; anthers caudate...........**Inuleae**

      I Pappus never capillary, often reduced or wanting; anthers not caudate.......**Anthemideae**

156

Key to the genera within the various tribes of the Compositae.

Heliantheae

  a Involucral bracts imbricated in 2 series, the inner membranous, very different
      from the outer in texture; ray flowers yellow or absent.

    b Inner bracts united to about the middle, the outer much smaller.............**Thelesperma** (nipple-weed)

    b Inner bracts distinct or slightly united at the base; ray flowers sometimes lacking.

      c Pappus of 2–5 downwardly barbed awns or teeth, outer bracts often
        as long or longer than the inner.................................**Bidens** (bur-marigold)

      c Pappus not downwardly barbed, the outer bracts very small.......................**Coreopsis** (tickseed)

  a Involucral bracts in 2 to several series, when only 2 series the inner and outer
      of similar texture.

    d Ray flowers pistillate, producing seeds.

      e Disk flowers also producing seeds; achenes not winged.............................**Heliopsis** (ox-eye)

      e Disk flowers not producing seeds, although they are perfect flowers;
        achenes of the ray flowers winged, imbricated in 2–3 rows.............**Silphium** (rosin-weed)

    d Ray flowers not producing seeds.

      f Corollas of ray flowers rose-purple; achenes not compressed
        or winged...**Echinacea** (purple cone flower)

      f Corollas of ray flowers yellow; achenes may be compressed and winged.

        g Achenes neither winged nor much compressed.

          h Receptacle flat, conic or convex; pappus of 2 large scales or
            awns, sometimes with 2–4 smaller ones.......................**Helianthus** (sunflower)

          h Receptacle conic or convex; pappus of very small teeth
            or absent....................................**Rudbeckia** (black-eyed susan)

        g Achenes compressed and winged; leaves pinrately divided.............**Ratibida** (cone-flower)

Helenieae

a Leaves decurrent on the stem; involucral bracts spreading or reflexed...........**Helenium (sneeze-weed)**
a Leaves not decurrent on the stem; involucral bracts appressed.................**Hymenoxys (hymenoxys)**

Anthemideae

a Leaves coarsely toothed; ray flowers one half inch or more long, white.....**Chrysanthemum (ox-eye-daisy)**
a Leaves very finely divided; ray flowers mostly under one fourth inch long, white or pinkish.................**Achillea (yarrow)**

Senecioneae

a Flowers white; ray flowers absent; sap milky; large veins striate............**Cacalia (Indian plantain)**
a Flowers yellow; ray flowers present; sap not milky; leaves various.................**Senecio (ragwort)**

Astereae

a Ray flowers, when present, yellow.
  b Pappus of scales or awns, sometimes deciduous, never of numerous capillary bristles; involucre viscid.................**Grindelia (gum weed)**
  b Pappus, at least in part, of numerous capillary bristles.
    c Pappus double, the inner of capillary bristles, the outer much shorter of scales or bristles; leaves not pinnatifid.................**Chrysopsis (golden-aster)**
    c Pappus wholly of capillary bristles.
      d Leaves pinnatifid.................**Haplopappus (golden-weed)**
      d Leaves entire or toothed, mostly lanceolate; ray flowers not more numerous than the disk flowers.................**Solidago (goldenrod)**
a Ray flowers blue, pink, purple or white, never yellow.

158

e Pappus a mere crown of a few scales or awn-like bristles..................Townsendia (townsendia)

e Pappus of numerous capillary bristles.

  f Bracts in 1–2 series, little imbricated; rays numerous, 20–150..................Erigeron (flea-bane)

  f Bracts in several series, the outer shorter, usually well imbricated.

    g Leaves entire or serrate, the teeth not bristle-tipped..................Aster (aster)

    g Leaves lobed; the lobes bristle-tipped..................Machaeranthera (viscid aster)

Inuleae

Pappus of capillary bristles; plants dioecious..................Antennaria (pussy-toes)

Eupatorieae

a Heads in broad cymes or panicles.

  b Pappus scabrous; achenes 5-angled..................Eupatorium (boneset)

  b Pappus plumose; achenes 8–10-striate..................Kuhnia (false boneset)

a Heads in elongated spikes or racemes; pappus often plumose..................Liatris (gay feather, blazing star)

Vernonieae

Coarse, erect perennial herbs with purple flowers..................Vernonia (ironweed)

Cynareae

a Pappus bristles plumose; leaves often sessile but not decurrent; heads purple, yellow or white..................Cirsium (thistle)

a Pappus bristles not plumose; leaves decurrent; heads purple..................Carduus (musk thistle)

Cichorieae

a Pappus of small scales, much shorter than the achenes; flowers blue; heads sessile
      or nearly so...Cichorium (chicory)
a Pappus bristles at least as long as the body of the achene.
   b Plants with leafy stems or with several heads.
      c Heads usually 5-flowered; flowers pink to lavender..........Lygodesmia (skeleton weed)
      c Heads usually many-flowered, or the flowers yellow.
         d Achene and pappus about 3 inches long.........................Tragopogon (goat's beard)
         d Achene and pappus much shorter.
            e Flowers blue...Lactuca (wild lettuce)
            e Flowers yellow; pappus soft, white....................................Crepis (hawk's beard)
   b Plants without stems; scapes 1-headed.
      f Leaves pinnatifid..Taraxacum (dandelion)
      f Leaves essentially entire...Microseris (false dandelion)

160

Glossary

Achene: a small, dry, nonopening fruit, having one cavity with one seed in it.

Acuminate: tapering to a slender point.

Actinomorphic: having radial symmetry as in a regular flower.

Acute: forming a sharp, definite angle at the base or apex.

Adnate: unlike parts grown together as stamens fused with the corolla tube.

Aerial: growing in air, above ground, or above water.

Alternate: borne at intervals at different levels; leaves which occur singly at a node.

Angled: having sharp ridges which are usually longitudinal; leaf segments which are pointed rather than rounded.

Annual: occurring every year; plants which grow from seeds each year.

Anther: part of a stamen containing pollen sacs in which pollen is formed.

Apomictic: having a type of apomixis which may involve any one of several variants from the usual sexual reproductive process; production of seeds without the usual process of fertilization.

Appressed: lying parallel to or flat against a surface.

Aquatic: growing in water.

Ascending: growing upward at an angle, neither prostrate nor erect.

Auricle: a basal ear-shaped part or lobe.

Awl-shaped: tapering from the base to a rigid point.

Awn: a terminal, stiff bristle.

161

Axil: the upper, usually acute, angle between an organ and its axis.

Axillary: in or from the axil.

Axis: the longitudinal support for several lateral organs; a stem or rachis.

Barbed: provided with lateral, usually reflexed, points.

Basal: located at the lowest part of an organ or at the base of the stem.

Beak: projection.

Bearded: possessing long or stiff hairs.

Berry: a kind of fleshy fruit which matures from a single ovary and contains few to many seeds.

Biennial: plants that live two years, producing vegetative growth the first year and a flower stalk the second.

Bilabiate: two-lipped.

Bipinnate: doubly pinnate.

Bisexual: having both stamens and pistils in a single flower.

Bladdery: inflated and having thin walls.

Blade: the thin and expanded portion of a leaf or petal.

Bloom: a waxy, usually whitish, coating of a stem, leaf, or fruit.

Bract: a small or scalelike leaf subtending a flower or a cluster of flowers in an inflorescence.

Bristle: a stiff hair.

Bud: a structure from which a branch or a flower develops.

Bulb: a subterranean, fleshy-leaved, erect bud capable of producing a new plant.

Calyx: the outer whorl of flower parts; the usually green outer parts of a flower bud; the group of sepals.

Campanulate: bell-shaped, or broadly cup-shaped, descriptive of a corolla or calyx.

Capillary: slender, hairlike.

Capitate: collected into a dense cluster; knoblike.

Capsule: a type of fruit which at maturity is dry and disperses many seeds. It arises from a compound ovary.

Carpel: a simple pistil or a unit of a compound pistil; a foliar unit, regarded as homologous to a leaf, which bears ovules.

Caudex: thick perennial base of a herbaceous plant.

Chaff: a small, thin, dry scale associated with the disk flowers of a composite.

Chlorophyll: a green photosynthetic pigment usually found in chloroplasts of the green tissue of stems, leaves, and sepals.

Ciliate: having a margin or edge lined with prominent stiff hairs.

Clasping: said of a structure such as a leaf base which lies close to and nearly surrounds the stem.

Claw: narrow stalk of a petal or sepal.

Cleft: deeply cut; often used to designate a cut about halfway. (More deeply cut would be called divided, and less deeply cut would be called lobed).

Cleistogamous: said of a flower that does not open.

Clone: plants which have been vegetatively propagated from one original.

Compound: having several parts.

Cordate: heart-shaped. Often used in reference to the shape of a leaf base with two rounded lobes and the sinus between.

Corm: a subterranean, erect, fleshy storage stem having thin, dry scales in place of leaves.

Corolla: the whole set of petals of one flower; the inner whorl of perianth parts.

Crenate: having a margin with rounded teeth.

Crest: a superior ridge.

Crown: an outgrowth of the throat of a corolla.

Cyme: a type of inflorescence bearing flowers at about the same height.

Cymose: resembling a cyme.

Decumbent: lying on the ground but with the tips ascending.

Decurrent: running down the stem from the place of insertion.

Decussate: opposite leaves with successive pairs at right angles to each other, forming four rows of leaves up the stem.

Dehiscence: the opening of an organ to liberate its contents, as of a capsule or an anther.

Deltoid: triangular in shape or merely having a broad base and an acute tip.

Dentate: having large, sharp teeth that are directed outward.

Dichotomous: dividing or forking into two divergent branches nearly equal in size.

Diffuse: widely and loosely spreading.

Digitate: compound with members arising at one place.

Dimorphic: of two forms.

Dioecious: unisexual flowers of two kinds borne on separate plants.

Discrete: separate; not united.

Disk: development of the receptacle around the base of the pistil; the center of a head type of inflorescence especially in the composite family.

Disk flower: the tubular flowers in the center of the head of most Compositae; in contrast to ray flowers.

Dissected: divided into many slender segments.

Distal: at the far end.

Distinct: separate; not united with similar parts.

Disturbed: said of areas in which the soil has been removed, cultivated, or compressed by wind, man, or animals.

Divergent: spreading broadly.

Divided: separated to the base. A simple leaf may be divided if the incisions from the margin reach nearly to the midrib.

Downy: descriptive of pubescence which consists of short, weak, soft hairs.

Drupe: a fleshy fruit having a hard or stony pit. The seed is inside the pit.

Elliptical: oval; elongated somewhat with rounded ends yet widest at the middle.

Emergent: plants or plant parts which grow into the aerial environment from an aquatic habitat. Parts which come above the surface of the water—not just floating on the surface.

Emersed: raised out of the water instead of floating on it.

Entire: without incisions, clefts, or sinuses; having no teeth, lobes, or divisions.

Erect: extending vertically or nearly so.

Exserted: growing beyond, as with stamens that grow out of a corolla tube.

Fascicle: a bundle.

Fertile: productive; stamens bearing pollen or fruits bearing seeds.

Filament: the stalk of the stamen which supports the anther; a threadlike structure.

164

Filiform: threadlike; long and slender.

Floret: a grass flower; a small flower, usually one of a number set closely together.

Follicle: a simple fruit, derived from a single carpel, opening along one line.

Forb: any herbaceous plant except grass.

Free: not attached to other organs. (Differs from distinct which signifies that the parts of one whorl are separate. If stamens are not joined to other parts such as the corolla they are said to be free. If stamens are separate from one another they are distinct).

Fruit: a mature and ripened ovary, sometimes with associated parts such as calyx or receptacle.

Fusiform: spindle-shaped; long and narrow but tapering at each end.

Genus: a group of closely related species.

Glabrate: becoming nearly hairless with age.

Glabrous: devoid of hairs or other pubescence.

Glandular: having glands—often only a single or a few cells at the end of a hair.

Glaucous: covered with a whitish bloom of waxy material which may be easily rubbed off.

Globose: nearly spherical.

Glomerate: clustered compactly.

Gynoecium: the female parts of a flower; the pistil or pistils of one flower.

Habit: the characteristic appearance of a plant.

Habitat: the characteristic place or environment in which a plant grows.

Hastate: like sagittate but with the basal lobes diverging.

Head: a type of flower cluster in which the stalkless florets are closely associated.

Herb: a plant having nonwoody stems above ground.

Herbaceous: dying at the end of the growing season.

Hirsute: having coarse, rough hairs.

Hispid: having bristly, stiff hairs.

Hoary: grayish-white because of the presence of fine hairs.

Hood: an appendage attached at the base of the gynoecium.

Horn: an appendage of the hood which is attached near the base of the hood.

Hybrid: the result of a cross between members of two species.

Incised: cut sharply and often irregularly.

Indehiscent: not opening at maturity, usually referring to dry fruits which are often one-seeded and do not scatter their seeds.

Indigenous: native to a given region.

Inferior: arising below another structure. (An ovary is inferior when it is sunken into the receptacle below the separation of the other parts of the flower.)

Inflorescence: the segment of a plant which bears flowers; a cluster of flowers.

Inserted: attached to (referring to the point of origin).

Internode: portion of an axis between the insertion of adjacent lateral organs; portion of the stem between two nodes.

Interrupted: not continuous.

Introduced: not native but brought into a given region.

Involucral: pertaining to an involucre.

Involucre: a set of bracts surrounding a flower or cluster of flowers.

Irregular: either asymmetrical or bilaterally symmetrical; not regular or radially symmetrical.

Keel: a projecting ridge resembling the keel of a boat; the two lower petals in many legume flowers.

Lamina: the thin portion of the blade of a leaf or petal, excluding the midrib.

Lanceolate: shaped like a lance; elongate and narrow, and widest below the middle.

Lateral: borne on the side of an axis.

Latex: the usually milky juice of certain plants which is produced in specialized cells or tubes.

Leaflet: a distinct part of a compound leaf.

Legume: a plant of the family Leguminosae such as a bean or pea; a simple dry fruit arising from a single carpel and dehiscing by two sutures as in a bean pod or pea pod.

Ligulate: strap-shaped.

Ligule: the straplike corolla of a ray flower in the Compositae; a flat outgrowth at the juncture of leaf blade and leaf sheath in grasses.

Limb: the outer, spreading free parts of a petal or sepal.

Linear: long and narrow; having nearly parallel margins.

Lip: one of the two projecting parts of an irregular calyx or corolla; the odd petal of an orchid.

Lobe: a segment of a larger structure somewhat set off by a space or sinus.

Locule: the cavity of an organ, such as an ovary or anther.

Loment: a legume fruit which is constricted between each seed.

Midrib: the thickened middle part of a leaf blade or leaflike organ.

Monoecious: having two kinds of unisexual flowers on the same plant.

Mucronate: having a small point at the end.

Naturalized: a nonindigenous plant, well established and maintaining itself without cultivation.

Nerve: a vascular bundle and associated strands of fibers which are thicker than other parts of the lamina of leaves, petals, bracts etc.

Node: a point or segment of a stem or axis from which a lateral organ arises.

Noxious: undesirable, pestiferous, or poisonous.

Obcordate: inversely heart-shaped; attached at the pointed end.

Oblanceolate: reverse of lanceolate in that the widest part is near the apex or tip.

Oblong: much longer than broad and having nearly parallel sides.

Obovate: reverse of ovate in having the broadest end near the apex or tip.

Obtuse: forming an angle greater than ninety degrees; having a blunt appearance.

Opposite: occurring in pairs at the same node directly across from each other; standing in front of an unlike organ.

Ovary: the basal, usually enlarged, part of the pistil which contains the ovules.

167

Ovate: having a shape similar to the profile of an egg with the broadest part near the base.

Ovule: the structure in which an egg is produced and fertilization takes place; the structure which becomes the seed.

Palmate: having the parts of a structure attached at a common point; same as digitate.

Panicle: repeatedly branched, often diffuse, type of flower arrangement.

Paniculate: like a panicle.

Pappus: a set of scales or bristles which replace the sepals in composite flowers.

Pedicel: the stalk of a single flower, when several flowers occur on one floral axis.

Peduncle: the stalk of a solitary flower.

Pendulous: hanging or drooping from its point of support.

Perennial: having parts which live from several to many years.

Perfect flower: a bisexual flower; one which has both stamens and pistil.

Perfoliate: having the leaves surrounding the stem.

Perianth: the nonessential parts of the flower; the calyx (sepals) and corolla (petals) collectively.

Persistent: remaining attached; not falling off.

Petal: a floral part, usually showy—those of one flower make up the corolla.

Petiole: the stalk of a leaf.

Petiolate: having a petiole.

Pilose: beset with soft spreading hairs.

Pinnate: feather-formed; having several similar parts disposed along a central mid structure.

Pinnatifid: pinnately divided or cleft.

Pistil: the central female part of the flower composed of a carpel or carpels and divided into stigma, style and ovary.

Pistillate: having pistils and implying a lack of stamens.

Plicate: folded lengthwise.

Plumose: plumy; having long fine hairs from a slender structure, such as the pappus bristles of some thistles.

Pod: capsule; a rather general uncritical term.

Pollen: a small particle containing male elements which later produces sperms necessary to fertilize the egg in an ovule.

Pollination: the act of transferring pollen from anther to stigma.

Pollinium: a sticky or waxy mass of pollen as produced in milkweed or orchid flowers.

Pome: a simple fleshy fruit such as an apple or pear which matures from an inferior ovary of several carpels.

Posterior: back side or the side next to the axis.

Prickle: a small, sharp outgrowth from the surface.

Procumbent: trailing or lying on the ground but not rooting.

Prostrate: lying flat on the ground; same as procumbent.

Pubescent: covered with short soft hairs; downy.

Punctate: having dotted or partially translucent areas.

Raceme: an arrangement of flowers, having each flower stalked on an unbranched axis, with the lowest flowers opening first.

Racemose: having flowers in racemes.

Rachis: the axis of a flower cluster or of a compound leaf.

Radiately: radiating from the center.

Ray: ligule; corolla of a ray flower.

Ray flower: one of the flowers of a head in the Compositae which has a straplike (ligulate) corolla.

Receptacle: the end of the axis which supports a flower and the structure from which separate flower parts arise; also the tissue from which the flowers of a head arise.

Recurved: turned downward, often with the outer surface on the inside of the curve.

Reflexed: bent backwards.

Regular: having parts of uniform size and shape, usually also implying radial symmetry (actinomorphic).

Reticulate: netlike, commonly used in regard to the vein pattern in a leaf.

Retrorse: directed toward the base or downward.

Rhizome: a slender, horizontal, underground stem.

Rib: a main vein of a leaf.

Rootstock: similar to rhizome but may include any underground structure which propagates the plant vegetatively.

Rosette: a closely clustered set of basal leaves.

Rotate: describes a set of united petals which is flat and nearly circular.

Rugose: wrinkled.

Runcinate: having margins which are sharply incised and have the segments directed toward the base rather than forward or straight out.

Runner: a stolon; an above-ground, horizontal stem capable of starting new plants by rooting at the nodes.

Sagittate: with the shape of an arrow-head, having narrow basal lobes.

Saprophytic: obtaining nourishment from dead organic matter.

Scabrous: rough to the touch.

Scale: a small, thin, flat, chaffy structure, not green.

Scape: a naked flower stalk arising from the ground.

Scarious: dry and membranous, not green.

Secund: attached along one side, or apparently so.

Segment: a part of an organ often set off by a space.

Sepal: one of the leaflike flower parts forming the outer whorl of a flower.

Sericeous: silky because of appressed white hairs.

Serrate: having sharp teeth pointing forward.

Sessile: without a supporting stalk.

Sheath: an envelope covering another organ.

Silique: a type of dry, dehiscent, simple fruit in which two outer halves of the ovary are shed leaving the seed attached to a thin partition.

Simple: usually refers to a single structure. (A simple leaf has a single blade. A simple pistil is composed of a single carpel. A simple fruit matures from a single pistil.)

Sinus: the incision or space between two lobes.

Smooth: not rough; not hairy.

Solitary: usually only one to a stem or one to a plant.

Spadix: flower-bearing, fleshy axis of a spike.

Spathe: a large solitary bract surrounding or subtending a spadix.

Spatulate: spoon-shaped.

Specific: definite; referring to a species.

Spike: an inflorescence which has an elongated axis bearing stalkless flowers.

Spine: a sharp outgrowth of a stem.

Spreading: extending over a larger area; becoming diffuse.

170

Spur: nectar-bearing tube of a corolla or calyx.

Squarrose: having tips recurved or at least spreading.

Stalk: any supporting axis.

Stamen: the pollen-producing organ of a flower, composed of a filament and an anther.

Staminate: having stamens; bearing pollen (with the implication that pistils are absent).

Standard: the upper petal in some legume flowers.

Stellate: starlike, said especially of hairs that have several arms arising from one point.

Stem: the axis of a shoot, usually bearing leaves at nodes.

Sterile: without reproductive parts in a flower.

Stigma: the glandular or receptive part of the pistil on which pollen grains germinate to produce pollen tubes.

Stipule: an appendage of the leaf base.

Stolon: a runner; an above-ground, horizontal, modified stem that may establish a new plant vegetatively.

Stoloniferous: bearing stolons.

Striae: longitudinal marks.

Striate: possessing longitudinal ribs, vascular bundles, lines, or stripes.

Strict: of erect habit and stiffly so.

Strigose: having appressed straight hairs.

Style: the narrowed upper part of the pistil which bears the stigma at its distal end.

Subtending: beneath.

Succulent: fleshy; soft; juicy.

Superior: describes an ovary that is free from the floral tube and the floral cup.

Symmetrical: a regular flower having the same number of floral parts in each whorl.

Taproot: the primary root which is a continuation of the lower end of the axis of the plant.

Tendril: a modified stem or leaf used for support.

Tepal: a showy flower part, used with monocot flowers that have similar-appearing sepals and petals.

Terminal: produced at or on the tip.

Ternate: divided into three main divisions.

Throat: the area near the junction of the tube and the limb of a corolla which has united petals.

Toothed: beset with outward-pointing projections.

Tribe: one of the subgroups of large families such as the Gramineae or the Compositae.

Trifoliate: bearing three leaves or three leaflets.

Truncate: cut off abruptly at the apex or base.

Tuber: a thickened, underground stem with many growing points.

Tubercle: a swollen area.

Turgid: swollen; tight; crisp.

Umbel: a type of inflorescence having the pedicels elongate and arising from nearly the same point.

Undulate: having a wavy margin or surface.

Unisexual: having stamens or pistils only (not both) in a flower.

Valve: one of the pieces of a dry dehiscent fruit.

Variants: a modification of the usual kind or color.

Vascular bundle: the set of conductive tissues making up a conductive pathway in a stem, petiole, or leaf blade; a vein.

Villous: bearing long, soft hairs; shaggy.

Viscid: sticky.

Weed: an unwanted plant; one growing out of place.

Whorl: a series of parts occurring at one level, usually around an axis.

Wing: a thin expanded part.

Woolly: covered with soft, long, entangled hairs; with a hairy covering matted together like wool.

Index

This index in a single alphabetical sequence contains common names (roman type), scientific names (italics), and family names (boldface).

174

178

179

180

Polygalaceae, 54
Polygonaceae, 16, 17
Polygonatum
  biflorum, 9
  commutatum, 9
Polygonum
  coccineum, 16
  longistylum, 16
  pensylvanicum, 16
  scandens, 17
pond lily, yellow, 23
Pontederiaceae, 6
poppy family, 28
poppy-mallow,
  pale, 56
  purple, 56
poppy, prickly, 28
Portulacaceae, 20
potato family, 91, 92
Potentilla
  anserina, 36
  arguta, 36
  recta, 37
prairie
  anemone, 24
  bean, 50
  cone-flower, 111
    gray-headed, 110
  dogbane, 72
  golden-aster, 117
  golden-pea, 50
  larkspur, 26
  milk vetch, 41
  mimosa, 42
  phlox, 82
  rose, 37
  shoe strings, 38
  snowball, 18
  sunflower, 108
  thermopsis, 50
  violet, 59
prairie-clover,
  purple, 47
  silky, 47
  white, 46
prairie-gentian, 71
prickly-pear, 61
prickly poppy, 28
primrose family, 69, 70
Primulaceae, 69, 70
Proboscidea louisiana, 98

Prunella vulgaris, 89
Psoralea
  argophylla, 48
  esculenta, 48
  tenuiflora, 49
puccoon,
  hairy, 84
  hoary, 84
  narrow-leaved, 84
puncture vine, 53
purple
  cactus, 60
  coneflower, 106
  poppy-mallow, 56
  prairie-clover, 47
purslane family, 20
pussy-toes, 122

Queen Anne's lace, 66

racemed milk-vetch, 40
ragwort, 115
Ranunculaceae, 24—27
Ranunculus
  aquatilis, 27
  circinatus, 27
  sceleratus, 27
Ratibida
  pinnata, 110
  columnaris, 111
rattleweed, 39
rayless thelesperma, 112
red clover, 50
red-seeded dandelion, 134
rocket, dame's, 32
Rocky Mountain bee plant, 30
Rorippa nasturtium-aquaticum, 33
Rosa
  arkansana, 37
  suffulta, 37
Rosaceae, 35—37
rose family, 35—37
rose,
  Arkansas, 37
  prairie, 37
  sunshine, 37
rosinweed, entire-leaved, 111
rough gay-feather, 123
Rubiaceae, 99
Rudbeckia
  hirta, 111

181

183